企业级卓越人才培养（信息类专业集群）解决方案"十三五"规划教材

SSH 轻量级框架实践

天津滨海迅腾科技集团有限公司 主编

南开大学出版社

天 津

图书在版编目(CIP)数据

SSH 轻量级框架实践 / 天津滨海迅腾科技集团有限公司主编 . — 天津：南开大学出版社，2017.5（2021.1 重印）
ISBN 978-7-310-05321-6

Ⅰ. ①S… Ⅱ. ①天… Ⅲ. ①JAVA语言—程序设计
Ⅳ. ①TP312.8

中国版本图书馆 CIP 数据核字（2017）第 015259 号

版权所有　侵权必究

SSH 轻量级框架实践
SSH QINGLIANGJI KUANGJIA SHIJIAN

南开大学出版社出版发行
出版人：陈　敬

地址：天津市南开区卫津路 94 号　邮政编码：300071
营销部电话：(022)23508339　营销部传真：(022)23508542
http://www.nkup.com.cn

唐山鼎瑞印刷有限公司印刷　全国各地新华书店经销
2017 年 5 月第 1 版　2021 年 1 月第 3 次印刷
260×185 毫米　16 开本　24.75 印张　620 千字
定价：58.00 元

如遇图书印装质量问题，请与本社营销部联系调换，电话：(022)23508339

企业级卓越人才培养（信息类专业集群）解决方案
"十三五"规划教材编写委员会

顾　问： 朱耀庭　南开大学
　　　　　 邓　蓓　天津中德应用技术大学
　　　　　 张景强　天津职业大学
　　　　　 郭红旗　天津软件行业协会
　　　　　 周　鹏　天津市工业和信息化委员会教育中心
　　　　　 邵荣强　天津滨海迅腾科技集团有限公司

主　任： 王新强　天津中德应用技术大学

副主任： 杜树宇　山东铝业职业学院
　　　　　 陈章侠　德州职业技术学院
　　　　　 郭长庚　许昌职业技术学院
　　　　　 周仲文　四川华新现代职业学院
　　　　　 宋国庆　天津电子信息职业技术学院
　　　　　 刘　胜　天津城市职业学院
　　　　　 郭思延　山西旅游职业学院
　　　　　 刘效东　山东轻工职业学院
　　　　　 孙光明　河北交通职业技术学院
　　　　　 廉新宇　唐山工业职业技术学院
　　　　　 张　燕　南开大学出版社有限公司

编　者： 王新强　李树真　吴　蓓　冯时昌　邬丕承

企业级卓越人才培养(信息类专业集群)
解决方案简介

　　企业级卓越人才培养(信息类专业集群)解决方案(以下简称"解决方案")是面向我国职业教育量身定制的应用型、技术技能型人才培养解决方案,以天津滨海迅腾科技集团技术研发为依托,联合国内职业教育领域相关行业、企业、职业院校共同研究与实践研发的科研成果。本解决方案坚持"创新产教融合协同育人,推进校企合作模式改革"的宗旨,消化吸收德国"双元制"应用型人才培养模式,深入践行"基于工作过程"的技术技能型人才培养,设立工程实践创新培养的企业化培养解决方案。在服务国家战略、京津冀教育协同发展、中国制造2025(工业信息化)等领域培养不同层次及领域的信息化人才。为推进我国教育现代化发挥应有的作用。

　　该解决方案由"初、中、高级工程师"三个阶段构成,集技能型人才培养方案、专业教程、课程标准、数字资源包(标准课程包、企业项目包)、考评体系、认证体系、教学管理体系、就业管理体系等于一体。采用校企融合、产学融合、师资融合的模式在高校内共建互联网学院、软件学院、工程师培养基地的方式,开展"卓越工程师培养计划",开设系列"卓越工程师班","将企业人才需求标准、企业工作流程、企业研发项目、企业考评体系、企业一线工程师、准职业人才培养体系、企业管理体系引进课堂",充分发挥校企双方特长,推动校企、校际合作,促进区域优质资源共建共享,实现卓越人才培养目标,达到企业人才培养及招录的标准。本解决方案已在全国近二十所高校开始实施,目前已形成企业、高校、学生三方共赢格局。未来五年将努力实现在年培养能力达到万人的目标。

　　天津滨海迅腾科技集团是以IT产业为主导的高科技企业集团,总部设立在北方经济中心——天津,子公司和分支机构遍布全国近20个省市,集团旗下的迅腾国际、迅腾科技、迅腾网络、迅腾生物、迅腾日化分属于IT教育、软件研发、互联网服务、生物科技、快速消费品五大产业模块,形成了以科技为原动力的现代科技服务产业链。集团先后荣获"全国双爱双评先进单位""天津市五一劳动奖状""天津市政府授予AAA级和谐企业""天津市文明单位""高新技术企业""骨干科技企业"等近百项殊荣。集团多年中自主研发天津市科技成果2项,具备自主知识产权的开发项目数十余项。现为国家工业和信息化部人才交流中心"全国信息化工程师"项目联合认证单位。

前　言

Java EE 如今已成为电信、金融、电子商务、保险、证券等行业大型应用系统的首选开发平台。在 Java EE 的开发中，SSH 轻量级框架占相当大的比重。

本书介绍的开发平台是以 Struts 2+Spring +Hibernate 为核心的轻量级 Java EE。书中对每个框架的知识点都进行了细化拆分，化整为零，使读者更加容易循序渐进地掌握 SSH 框架的核心技术。

本书以 SSH 框架的项目开发过程为线索，分为十一章，分别讲解了 Java EE 的概念和基础知识、Ajax 在 Web 开发中的作用和使用方法、Hibernate 的核心配置和映射、Spring 中 IOC、DI 和 AOP 的基本概念和应用、SSH 框架的整合等。

通过本书的学习，能够使读者分别掌握每个框架所讲的知识和在项目开发中的作用，还能够使读者了解三个框架在项目开发中的联系，让读者能够熟练使用 SSH 开发完整的 Java Web 项目。

本书每章均划分为学习目标、课前准备、本章简介、具体知识点讲解、小结、英语角、作业、思考题、读者回顾内容九个模块。学习目标和课前准备对本章要讲解的知识进行了简述，小结部分对本章知识进行了总结，英语角解释了本章一些术语的含义，可以使读者全面掌握本章所讲的内容。

本书由王新强、吴蓓、李树真、冯时昌、邬丕承共同编写，王新强担任主编工作，负责全书内容设计，吴蓓负责内容规划、编排。具体分工如下：第一、二、三章由王新强、吴蓓共同编写，第三、四、五章由李树真、冯时昌共同编写，第六至十一章由王新强、邬丕承、冯时昌共同编写。

本书不局限于对枯燥理论知识的讲解，而是采用理论和实战相结合的方式来讲授知识点，通过项目实战的练习使读者对理论知识熟练掌握。本书的每个实例之后都有相应的运行结果图，能使读者更加直观地看到每个示例的效果，理解实例所讲的知识和讲解的思路。

目 录

理论部分

第1章 Java EE 概述 ·········· 3
1.1 Java EE 概述 ·········· 4
1.2 应用服务器 ·········· 9
1.3 市场主流的应用服务器 ·········· 10
1.4 EJB ·········· 11
1.5 Java EE 优势 ·········· 12
1.6 Java EE 主要技术 ·········· 13
1.7 安装 WebLogic 服务器 ·········· 17
1.8 部署 Web 应用程序 ·········· 19
1.9 小结 ·········· 20
1.10 英语角 ·········· 20
1.11 作业 ·········· 20
1.12 思考题 ·········· 20
1.13 学员回顾内容 ·········· 20

第2章 使用 Ajax 改进用户体验 ·········· 21
2.1 Web 3.0 ·········· 21
2.2 Ajax ·········· 23
2.3 Ajax 技术结构 ·········· 35
2.4 信息搜索示例 ·········· 40
2.5 小结 ·········· 47
2.6 英语角 ·········· 48
2.7 作业 ·········· 48
2.8 思考题 ·········· 48
2.9 学员回顾内容 ·········· 49

第3章 Ajax 实战技巧 ·········· 50
3.1 完成表单验证 ·········· 50
3.2 动态加载列表框 ·········· 55
3.3 实现自动完成 ·········· 63
3.4 小结 ·········· 70

3.5	英语角	71
3.6	作业	71
3.7	思考题	71
3.8	学员回顾内容	71

第 4 章 使用 Hibernate 完成对象持久化 …… 72

4.1	为什么需要 Hibernate	73
4.2	Hibernate 是什么	74
4.3	使用 Hibernate 的基本步骤	76
4.4	使用 Hibernate 实现数据的删改	86
4.5	使用工具简化 Hibernate 开发	90
4.6	在项目中使用 Hibernate	93
4.7	小结	97
4.8	英语角	98
4.9	作业	98
4.10	思考题	98
4.11	学员回顾内容	99

第 5 章 Hibernate 的关联映射 …… 100

5.1	关联	100
5.2	多对多关联	118
5.3	小结	124
5.4	英语角	125
5.5	作业	125
5.6	思考题	125
5.7	学员回顾内容	125

第 6 章 Hibernate 查询 …… 126

6.1	Hibernate 查询语言	126
6.2	Criteria 查询	137
6.3	小结	139
6.4	英语角	140
6.5	作业	140
6.6	思考题	140
6.7	学员回顾内容	140

第 7 章 Hibernate 高级特性 …… 141

7.1	Hibernate 检索策略	141
7.2	Hibernate 缓存	150
7.3	小结	155

- 7.4 英语角 ... 156
- 7.5 作业 ... 156
- 7.6 思考题 ... 156
- 7.7 学员回顾内容 ... 156

第 8 章 Spring 原理与应用 ... 157

- 8.1 Spring 概述 ... 158
- 8.2 Spring 核心技术 ... 161
- 8.3 Spring 中的 Bean ... 165
- 8.4 依赖注入 ... 172
- 8.5 小结 ... 184
- 8.6 英语角 ... 184
- 8.7 作业 ... 185
- 8.8 思考题 ... 185
- 8.9 学员回顾内容 ... 185

第 9 章 Spring DI 详解 ... 186

- 9.1 Spring DI 机制的基本原理 ... 186
- 9.2 注解方式注入依赖对象 ... 197
- 9.3 小结 ... 202
- 9.4 英语角 ... 203
- 9.5 作业 ... 203
- 9.6 思考题 ... 203
- 9.7 学员回顾内容 ... 203

第 10 章 Spring AOP ... 204

- 10.1 面向方面编程概述 ... 204
- 10.2 AOP 详解 ... 211
- 10.3 Spring+JDBC 组合开发 ... 217
- 10.4 小结 ... 221
- 10.5 英语角 ... 222
- 10.6 作业 ... 222
- 10.7 思考题 ... 222
- 10.8 学员回顾内容 ... 222

第 11 章 Spring 与 Struts、Hibernate 的集成 ... 223

- 11.1 Spring 与 Hibernate 的集成 ... 223
- 11.2 使用 Spring 重新组装 Web 程序 ... 228
- 11.3 声明式事务 ... 235
- 11.4 小结 ... 244

11.5	英语角	245
11.6	作业	245
11.7	思考题	245
11.8	学员回顾内容	245

上机部分

第1章 Java EE 概述 ········· 249

1.1	指导（1 小时 10 分钟）	249
1.2	练习	249
1.3	作业	259

第2章 使用 Ajax 改进用户体验 ········· 260

2.1	指导（1 小时 10 分钟）	260
2.2	Ajax 的实现方式完成用户登录	260
2.3	作业	270

第3章 Ajax 实战技巧 ········· 271

3.1	指导（1 小时 10 分钟）	271
3.2	练习	289
3.3	作业	290

第4章 使用 Hibernate 完成对象持久化 ········· 291

4.1	指导	291
4.2	练习	295
4.3	作业	302

第5章 Hibernate 的关联映射 ········· 303

5.1	指导	303
5.2	练习	303
5.3	作业	316

第6章 Hibernate 查询 ········· 318

6.1	指导	318
6.2	练习	318
6.3	作业	328

第7章 Hibernate 高级特性 ········· 329

| 7.1 | 指导 | 329 |
| 7.2 | 练习 | 329 |

7.3 作业 339

第 8 章 Spring 原理与应用 340
8.1 指导 340
8.2 练习 340
8.3 作业 348

第 9 章 Spring DI 详解 349
9.1 指导 349
9.2 练习 349
9.3 作业 356

第 10 章 Spring AOP 357
10.1 指导 357
10.2 练习 357
10.3 作业 360

第 11 章 Spring 与 Struts、Hibernate 的集成 363
11.1 指导 363
11.2 练习 363
11.3 作业 379

理论部分

第 1 章　Java EE 概述

学习目标

- ◆ 了解什么是多层结构。
- ◆ 了解什么是 Java EE 容器。
- ◆ 熟悉常用的应用服务器。
- ◆ 了解 Java EE 的优势。
- ◆ 能够安装 WebLogic 服务器。

课前准备

熟悉常用的应用服务器。

本章简介

这一章介绍了 Java EE 的基本概念、Java EE 核心的 API 和 Java EE 中不同的开发角色。Java EE 技术提供了一整套基于组件的方案来设计、开发、装配和部署企业级应用程序，为开发者开发企业级的解决方案提供了强有力的支持。本章的重点是 WebLogic 服务器的安装、配置和相应的环境变量的修改。

今天，越来越多的开发者想要编写企业级分布式事务处理的应用程序，而这些应用程序必须要在速度、安全性和可靠性等方面发挥出色。如果你已经在这一领域从事工作，你应该了解在当今这个技术高速发展、要求苛刻的电子商务和信息技术的世界里，企业级的应用程序必须具有以下特点：花费更少的金钱、具有更快的开发速度和占用更少的资源。

为了减少费用，快速设计和开发企业级的应用程序，Java EE 平台提供了一个多层结构的分布式应用程序模型，该模型具有重用组件的能力、基于扩展标记语言（XML）的数据交换、统一的安全模式和灵活的事务控制。你不仅可以比以前更快地开发出新的解决方案，而且独立于平台的基于组件的 Java EE 解决方案不再受任何软件服务器厂商和应用程序编程接口（API）的限制。客户可以自己选择最合适于他们的商业应用和所需技术的产品和组件。

1.1 Java EE 概述

1.1.1 Client/Server 结构

当今,许多企业都需要扩展他们的业务范围,降低自身经营成本,缩短与客户之间的响应时间,这就需要一种简捷快速服务于企业、合作伙伴和雇员之间的应用软件。一般来说,提供这些服务的应用软件必须同企业信息系统(EIS)相结合,并提供新的服务。这些服务要具备以下的特点:

- 高可用性:以满足现在的全球商业环境。
- 安全性:保护用户的隐私和企业数据的安全。
- 可依赖性和可扩展性:保证商业交易的正确和迅捷。

最初这些服务是由两层应用(也称为客户/服务器或 C/S 结构)来实现的。其结构如图 1-1 所示。在这种结构下,服务器只提供唯一的服务,即数据库服务。客户端负责数据访问,业务逻辑处理,并将结果转换为某个形式以便向用户显示,以及接受用户的输出。客户/服务器的体系在开始的时候很容易配置,但难以升级或者扩展,而且通常基于私有的协议(典型的是私有的数据库协议)。在 Web 领域中,最重要的就是扩展,而两层的应用不便于升级扩展,因此并不适合用在广域网中。

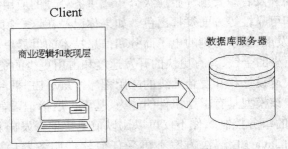

图 1-1 Client/Server 结构示意图

为了解决两层体系的不足,在客户端和后端数据源之间创建中间层,中间层提供了把商业功能和数据与 EIS 相结合的功能,它们把客户端从复杂的业务逻辑中分离出来,利用成熟的互联网技术使用户在管理上所花费的时间最小化。而 Java EE 正是降低了开发这种中间层服务的成本和复杂程度,因而使得服务可以快速展开,并能够更轻松地面对竞争中的压力。

1.1.2 多层结构

Java EE 平台使用了多层分布式的应用程序模型。根据其实现的不同功能被封装到不同的组件中，构成 Java EE 应用程序的组件。根据在其所属的不同层，被安装到不同的机器中。图 1-2 表示了多层的 Java EE 应用程序。在图中涉及的 Java EE 应用程序的各个部分将在 Java EE 组件中给出详细描述。

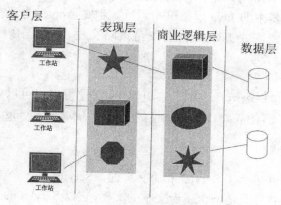

图 1-2　多层结构示意图

- 运行在客户端机器的客户层组件。
- 运行在应用服务器中的 Web 层组件。
- 运行在应用服务器中的商业逻辑层组件。
- 运行在 EIS 服务器中的企业信息系统（EIS）层软件。

从图中可以看到，Java EE 应用程序既可以是三层结构，也可以是四层结构，但是我们通常将 Java EE 应用程序的多层结构考虑为三层结构。这是因为它们分布在三个不同的位置：客户端机器、应用服务器机器和在后端的传统的大型机。三层结构的应用程序可以理解为在标准的两层结构的客户端应用程序和后端存储资源中间增加了一个多线程的应用服务器。

一、Java EE 组件

Java EE 应用程序由组件组成。一个 Java EE 组件就是一个带有特定功能的软件单元随同与它相关的类和文件被装配到 Java EE 应用程序中，并实现了与其他组件的通信。Java EE 规范中是这样定义 Java EE 组件的：

- 客户端应用程序和 Applet 是运行在客户端的组件。
- Java Servlet 和 Java Server Pages（JSP）是运行在服务器端的 Web 组件。
- Enterprise JavaBean（EJB）组件是运行在服务器端的商业逻辑组件。

Java EE 组件由 Java 编程语言写成，并和用 Java 写成的其他程序一样进行编译。Java EE 组件和其他的 Java 类的不同点在于：它存在于一个 Java EE 应用程序中，具有固定的格式并遵守 Java EE 的规范，它被部署在应用服务器中，应用服务器对其进行管理。

一个 Java EE 客户端既可以是一个 Web 客户端，也可以是一个应用程序客户端。

1. Web 客户端

一个 Web 客户端由两部分组成：运行在 Web 层的由 Web 组件生成的包含各种标记语言

（HTML、XML 等）动态 Web 页面和接收从服务器传送来并将它显示出来的 Web 页面。

一个 Web 客户端有时被称为瘦客户端。瘦客户端一般不做像数据库查询、执行复杂的商业规则及连接传统应用程序这样的操作。当你使用一个瘦客户端时，像这样重量级的操作被交给了在应用服务器中运行的 EJB。这样就可以充分发挥应用服务器技术在安全性、速度、耐用性和可靠性方面的优势。

Applet 也可以作为 Web 客户端，Applet 是一个用 Java 编程语言编写的小的客户端应用程序，它安装在 Web 浏览器中的 Java 虚拟机中运行。但是 Applet 为了在 Web 浏览器中成功地运行，客户端系统很可能需要 Java 插件和安全策略文件，所以现在已经很少把 Applet 作为 Web 客户端来使用了。

2. 应用程序客户端

Java EE 应用程序客户端运行在客户端机器上，它使得用户可以处理比 HTML 页面所能提供的更丰富的用户界面任务。具有代表性的是使用 Java 的 Swing 或 Awt 建立的图形用户界面（GUI）。

二、Web 组件

Java EE 的 Web 组件包括了 Servlet 和 JSP 页面。Servlet 是一个 Java 类，它可以动态地处理请求并做出响应。JSP 页面是一个基于文本的文档，它以 Servlet 的方式执行，但是它可以更方便地建立静态内容。

三、商业逻辑组件

商业逻辑代码，表示与银行、零售和航空等特定的商业领域相适应的商业逻辑。它由运行在商业逻辑层的 EJB 进行处理。

四、企业信息系统层

企业信息系统层处理企业信息系统软件，包含诸如企业资源计划（ERP）、主机事务处理、数据库系统和其他传统系统这样的底层系统。Java EE 应用程序可以访问企业信息系统。例如：访问数据库。

1.1.3 Java EE 容器

通常多层应用程序是很难编写的，因为这包括许多复杂的代码如处理事务、状态管理、多线程、组件池和其他复杂的底层。基于组件与平台无关的 Java EE 体系结构，使得 Java EE 应用程序易于编写，这是因为商业逻辑被封装到可重用的组件中。此外，Java EE 应用服务器以容器的形式为每一个组件类型提供底层服务。因此我们不需要自己开发这些服务，这使我们可以全力以赴地着手处理商业逻辑问题。

一个 Java EE 应用程序运行前，它们必须被部署到 Java EE 应用服务器中。通常，一个 Java EE 应用服务器提供了 EJB 容器和 Web 容器。Enterprise JavaBean（EJB）容器用来管理 Java EE 应用程序的 EJB 的运行。Web 容器用来管理 Java EE 应用程序中 JSP 页面和 Servlet 等 Web 组件的运行。

客户端可以直接和运行在 Java EE 服务器中的业务逻辑层进行通信，如果是一个运行在浏览器中的客户端，也可以通过运行在 Web 层中的 JSP 页面或 Servlet 与其进行通信。

业务逻辑代码表示与特定商业领域（例如银行、零售等行业）相适应的逻辑。它由运行在业务逻辑层的 Enterprise Bean 处理。一个 Enterprise Bean 可以从客户端接收数据，对它进行处理，并将其发送到企业信息系统层以作存储；同时它也可以从存储器获取数据，处理后将其发送到客户端应用程序。有三种类型的 Enterprise Bean：Session Bean、Entity Bean 和 Message-Driven Bean。Session Bean 描述了与客户端的一个短暂会话。当客户端执行完成后，Session Bean 和它的数据都将消失；与之相对应的是一个 Entity Bean，描述了存储在数据库表中的一行持久稳固的数据，如果客户端终止或者服务结束，底层的服务会负责 Entity Bean 数据的存储。Message-Driven Bean 结合了 Session Bean 和 Java 信息服务（JMS）信息监听者的功能，它允许一个商业组件异步地接受 JMS 消息。

容器是一个组件和支持组件的底层平台特定功能之间的接口，在一个 Web 组件、Enterprise Bean 或者是一个应用程序客户端组件被执行前，它们必须被装配到一个 Java EE 应用程序中，并且部署到容器中。装配的过程包括为 Java EE 应用程序中的每一个组件以及 Java EE 应用程序本身指定容器的设置。容器设置定制了由 Java EE 服务器提供的底层支持，这将包括安全性、事务管理、Java 命名目录接口（JNDI）搜寻以及远程连接等。下面是其中的主要部分：

● Java EE 的安全性模式可以让开发者对一个 Web 组件或 Enterprise Bean 进行配置使得只有授权用户可以访问系统资源。

● Java EE 的事务模式可以让开发者指定方法之间的关系组成单个的事务，这样在一个事务中的所有方法将被视为单一的整体。

● JNDI 搜寻服务为企业中的多种命名目录服务提供一个统一的接口，这使得应用程序组件可以访问命名目录服务。

● Java EE 远程连接模式管理客户端和 Enterprise Bean 之间的底层通信。在一个 Enterprise Bean 被建立后，客户端在调用其中的方法时就像与这个 Enterprise Bean 直接运行在同一个虚拟机上一样。

实际上，Java EE 体系结构提供的可配置的服务意味着在相同的 Java EE 应用程序中的应用程序组件根据其被部署在什么地方而在实际运行时会有所不同。例如，一个 Enterprise Bean 可能在一个产品环境中拥有包含访问数据库数据的某种级别的安全性设置，而在另一个产品环境中是另一个访问数据库的级别。

容器还管理诸如 Enterprise Bean 和 Servlet 的生存周期、数据库连接资源池以及访问在 Java EE API 中介绍的 Java EE 平台 API 这样不能配置的服务。尽管数据持久化是一个不能配置的服务，但是 Java EE 体系结构允许你在想要获得比默认的容器管理持久化所能提供的更多的控制时，通过在你的 Enterprise Bean 执行中包含适当的代码以重载容器管理持久化。例如，你可以使用 Bean 管理持久化以实现你自己的 finder（查找）方法或者是建立一个定制的数据库缓冲区。

Enterprise JavaBean（EJB）容器：EJB 容器是 Java EE 服务器的一部分。Java EE 服务器是 Java EE 产品的运行部分，一个 Java EE 服务器提供 EJB 容器和 Web 容器。EJB 容器负责管理 Java EE 应用程序中 Enterprise Bean 的执行，Enterprise Bean 和它的容器运行在 Java EE 服务器中。

Web 容器：Web 容器管理 Java EE 应用程序的 JSP 页面和 Servlet 组件的执行，Web 组件和它的容器也运行在 Java EE 服务器中。

1.1.4 开发角色

可重用的组件将应用程序开发过程和部署过程分成不同的角色成为可能。部署就是指通过适当的设置让应用程序在服务器中运行起来。这样，不同的人或公司可以在开发过程和部署过程的各个部分承担不同的任务。如果按照不同任务划分，可以划分成以下不同的角色。

1. Java EE 产品提供者

Java EE 产品提供者是设计并提供实现 Java EE 规范所定义的 Java EE 平台、API 和其他功能的公司。这一般是指操作系统、数据库系统、应用服务器或 Web 服务器的厂商，它们依照 Java EE 的规范实现 Java EE 平台。

2. 工具提供者

工具提供者是指建立组件提供者、装配者和部署者所使用的开发、装配和打包的工具的公司或个人。例如工具提供者可以提供符合 Java EE 规范的管理和监视应用服务器运行情况的工具。

3. 应用程序组件提供者

应用程序组件提供者是指建立 Java EE 应用程序所使用的 Web 组件、EJB、Applet 或应用程序客户端的公司或个人。

4. 应用程序装配者

应用程序装配者是从应用程序组件提供者接受应用程序组件的 JAR 文件并将其组装成一个 Java EE 应用程序 EAR 文件的公司或个人。装配者或部署者可以直接编辑部署描述文件或者使用交互式的工具，正确地添加 XML 标志。

5. 应用程序部署者和系统管理员

应用程序部署者和系统管理员应该是两种角色，但是在实际开发的过程中，他们的任务是由同一个角色担当。应用程序部署者和系统管理员可以是公司或个人，他们配置和部署 Java EE 应用程序，管理 Java EE 应用程序在其中运行的计算机和网络这些底层结构，并对运行环境进行监控。他们的任务可能包括这样的一些事：设置事务控制、安全属性并指定数据库连接。

在配置时，部署者按照由应用程序组件提供者提供的指示以解决外部的支持、指定安全设置并定义事务属性。在安装时，部署者将应用程序组件装入服务器中并生成特定容器的类和接口。

6. 系统组件提供者

系统组件提供者提供连接和访问现存的企业信息系统（EIS）所需要的工具和主要的 API，企业信息系统（EIS）通常指 ERP、数据库和消息系统。

通常由系统管理员承担购买和安装 Java EE 产品和工具的任务。在购买和安装好软件之后，应用程序组件提供者开发 Java EE 组件，应用程序装配者负责装配，而应用程序部署者负责部署。在一个大的组织中，每个这样的角色可能对应于不同的个人和小组。做这样的分工是因为前一个角色将会输出一个可移植的文件，而这将是后一个角色的输入。例如，在应用程序组件提供阶段，应用程序组件提供者提交 EJB JAR 文件。而在应用程序装配阶段，另一些开发者将这些 EJB JAR 文件组合到一个 Java EE 应用程序中并将其保存为一个 EAR 文件。在应用程序部署阶段，系统管理员将这个 EAR 文件安装到 Java EE 应用服务器中。

不同的角色不一定总得由不同的人来执行。如果你在一个小公司工作或者你是一个自由程序员，可能你需要执行每一个阶段的任务。

1.2 应用服务器

Java EE 应用服务器以容器的形式为每一个组件类型提供底层服务。因为我们不需要自己开发这些服务，这使我们可以全力以赴地着手处理商业逻辑问题。

一个 Java EE 应用程序运行前，必须被部署到 Java EE 应用服务器中。通常，一个 Java EE 的应用服务器提供了 EJB 容器和 Web 容器。Enterprise JavaBean（EJB）容器用来管理 Java EE 应用程序的 EJB 的运行。Web 容器用来管理 Java EE 应用程序中 JSP 页面和 Servlet 等 Web 组件的运行。

主要的应用服务器厂商

应用服务器主要用于提高开发人员的生产力，减轻管理的复杂度，为集成应用提供强化服务。应用服务器的市场得到迅速的发展，下面列举了在应用服务器市场上占有重要份额的应用服务器厂商和他们的应用服务器产品。

1. BEA WebLogic

BEA WebLogic 应用服务器是应用服务器市场上占有率最高的应用服务器产品，使用广泛，电子商务中最著名的 Amazon.com 所采用的就是 BEA 的 WebLogic 服务器。本书中的例子程序所使用的应用服务器就是 BEA WebLogic 服务器。更多的详细资料请参阅其主页 http://www.bea.com。

2. IBM WebSphere

IBM WebSphere 应用服务器是另一款非常流行的应用服务器产品。WebSphere 在中国的应用还是很广泛的，针对 WebSphere，IBM 在中国不但举办了很多的竞赛和讲座，还与国内诸多高校合作，免费赠送了很多套 WebSphere 的套件。更多的详细资料请参阅其主页 http://www-4.ibm.com/software/webservers/appserv/。

3. Sun ONE Application Server

Sun ONE Application Server 是 Sun 公司最新推出的应用服务器产品，最大的特点是 100%符合 Java EE 规范。请参阅其主页 http://wwws.sun.com/software/products/appsrvr/home_appsrvr.html。

4. Borland Enterprise Server，AppServer Edition

Borland Enterprise Server，AppServer Edition 是 Borland 公司推出的应用服务器产品，最大的特点是 JBuider 对 AppServer 的支持非常好，毕竟 JBuider 和 AppServer 都是 Borland 公司的产品。更多的详细资料请参阅其主页 http://www.borland.com/besappserver/index.html。

5. JBOSS

JBOSS 的最大特点是开放源代码的应用服务器产品。更多的详细资料请参阅其主页 http://www.jboss.org。

1.3 市场主流的应用服务器

1.3.1 WebLogic 简介

BEA WebLogic 是用于开发、集成、部署和管理大型分布式 Web 应用、网络应用和数据库应用的 Java 应用服务器。将 Java 的动态功能和 Java Enterprise 标准的安全性引入大型网络应用的开发、集成、部署和管理之中。

BEA WebLogic Server 拥有处理关键 Web 应用系统问题所需的性能、可扩展性和高可用性。BEA WebLogic Server 可为部署个性化电子商务应用系统提供完善的解决方案。BEA WebLogic Server 具有开发和部署关键任务电子商务 Web 应用系统所需的多种特色和优势，主要包括以下几点：

- 领先的标准

对业内多种标准的全面支持，包括 EJB、JSB、JMS、JDBC、XML 和 WML，使 Web 应用系统的操作更为简单，并且保护了投资，同时也使基于标准的解决方案的开发更加简便。

- 无限的可扩展性

BEA WebLogic Server 以其高扩展的架构体系闻名于业内，包括客户机连接的共享、资源 pooling 以及动态网页和 EJB 组件群集。

- 快速开发

凭借对 EJB 和 JSP 的支持，以及 BEA WebLogic Server 的 Servlet 组件架构体系，可加快投放市场速度。这些开放性标准与 JBuilder 配合时，可简化开发，并可发挥已有的技能，迅速部署应用系统。

- 部署更趋灵活

BEA WebLogic Server 的特点是与数据库、操作系统和 Web 服务器紧密集成。

- 关键任务可靠性

其容错性、系统管理和安全性能已经在全球数以千计的关键任务环境中得以验证。

BEA WebLogic Server 是专门为企业电子商务应用系统开发的。企业电子商务应用系统需要快速开发，并要求服务器端组件具有良好的灵活性和安全性，同时还要支持关键任务所必需的扩展性和高可用性。BEA WebLogic Server 简化了可移植及可扩展的应用系统开发，并为其他应用系统提供了丰富的互操作性。

凭借其出色的群集技术，BEA WebLogic Server 拥有最高水平的可扩展性和可用性。BEA WebLogic Server 既实现了网页群集，也实现了 EJB 组件群集，而且不需要任何专门的硬件或操作系统支持。

无论是网页群集，还是组件群集，对于电子商务解决方案所要求的可扩展性和可用性都是至关重要的。

1.3.2 JBOSS 介绍

JBOSS 是开放源代码的，遵从 Java EE 规范的，100%纯 Java 的 EJB 服务器。JBOSS 的开发团队有 500 多人，核心开发人员有 50 多个，均为 Java EE 专家。JBOSS 的用户数以万计，遍及世界。JBOSS 采用 Java Management Extension API 实现软件模块的集成与管理。 JBOSS 的详细资料请参阅其主页 http://www.jboss.org。

1.3.3 WebSphere 介绍

WebSphere 应用服务器是一个完善、开放的应用服务器，是 IBM E-Business 应用架构的核心。WebSphere 应用服务器提供了开放、标准的平台和工具，以加速向网上服务的迁移。WebSphere 应用服务器满足维护一个 Web 站点的任何需求：安装简单，图形用户界面（方便 Servlet 管理），基于 Web 的远程管理和安全特性。

1.4 EJB

Java EE 技术赢得广泛重视的主要原因之一就是 EJB（Enterprise JavaBean），要注意的是，EJB 从技术上而言不是一种产品，而是一个技术规范。它提供了一个框架来开发和实施分布式商务逻辑，由此简化了具有可伸缩性和高度复杂的企业级应用的开发；此外，它还定义了 EJB 组件在何时、如何与它们的容器进行交互作用，容器负责提供公用的服务，例如目录服务、事务管理、安全性、资源缓冲池以及容错性。有关 EJB 框架的 API 定义在 Javax.EJB 及其子包 javax.ejb.spi 中。

为了满足架构的目标，EJB 规范中定义了如下一些新的概念：

● EJB 服务器（EJB Server）负责管理 EJB 容器（它负责管理 Beans），提供对操作系统服务的存取和 Java 相关的服务，尤其是通过 JNDI 访问命名空间和基于 OTS 的事务处理服务。

● EJB 容器（EJB Container）负责管理部署于其中的 Enterprise Bean。客户机应用程序并不直接与 Enterprise Bean 进行交互。相反，客户机应用程序通过由容器生成的两个封装接口与 Enterprise Bean 进行交互。当客户机使用封装接口调用各种操作时，容器截获每个方法调用，并插入管理服务。对客户端而言 EJB 容器是透明的。

● EJB 客户端（EJB Client）可以是 Servlet、JSP、应用程序或其他 Bean。客户端可以通过 JNDI 来查找 EJB Home 接口，步骤如下：首先创建一个 JNDI Context（initial context）；然后使用 JNDI Context 来查找 EJB Home 接口；再使用 EJB Home 接口来创建或查找 EJB 实例；最后使用 EJB 实例完成业务操作。注意实际的存取（对 EJB）是通过容器生成的类来完成的。

每个 Enterprise JavaBean 都必须实现 Home 接口（扩展自 EJB Home 接口）和 Remote 接口（扩展自 EJB Object 接口），简单来说，Home 接口定义了如何查找或创建 Bean，Remote 接口定义了 Bean 的业务方法。

1.5 Java EE 优势

在两层体系下，业务逻辑在客户端和服务器端都有分布，而且服务器端的主要计算任务都由数据库承担。数据库系统本身作为系统级软件，它的优势在于提供高效率的数据访问和处理而不是数值计算，利用数据库系统处理大量的业务计算并没有完全发挥出数据库管理系统的特长，当请求并发量巨大时，数据库性能下降很快。针对这一不足，三层架构的处理方式是：业务逻辑分布到应用服务器上，数据库上不再具有业务逻辑处理单元，而只负责基础业务数据的管理，主要的计算任务由应用服务器完成，从而充分利用了应用服务器在并发处理和逻辑计算方面的优势。另外，应用服务器还可以做集群的配置，即在物理上，统一应用管理多台应用服务器对外部请求的分配和并行处理。这样，当计算请求并发量巨大时，集群的多台应用服务器之间可以动态地进行任务分配，实现负载均衡，保证了系统性能不会因为大量并发用户的访问而急剧下降。另外，系统也同时具备了很好的可扩展性和伸缩性，即在请求并发量增大或减少时，可根据实际情况增加或减少应用服务器的数量，以便在保证性能的前提下，合理利用硬件资源。这些都是两层结构所不具备的优点。

基于 Java EE 的三层架构还具备很多特点。应用 Java EE 的三层架构对于保留现存的 IT 资产有着重要意义。在信息系统建设方面，做平滑的系统迁移方案，而不是重新制定全盘方案就变得很重要。这样，一个以渐进的，而不是激进的、全盘否定的方式建立在已有系统之上的服务器端平台机制，Java EE 架构可以充分利用用户原有的投资，如一些公司使用的 BEA Tuxedo、IBM CICS、IBM Encina、Inprise VisiBroker。之所以成为可能，主要是因为 Java EE 拥有广泛的业界支持。每一个供应商都对现有的客户提供了不用废弃已有投资，进入可移植的 Java EE 领域的升级途径。另外，由于基于 Java EE 平台的产品几乎能够在任何操作系统和硬件配置上运行，现有的操作系统和硬件也可以被完整地保留下来继续使用。

Java EE 体系对大型系统的高效开发有着良好的支持。Java EE 标准严格要求把一些通用的、很繁琐的服务端底层开发任务交给中间件供应商去完成，而这些复杂的系统级功能是常规应用开发中难度最大、开发成本最高的一部分工作。高级中间件供应商提供复杂的中间件服务，如：状态管理服务、持续性服务、分布式共享数据对象的缓冲服务等，它对开发人员来说是很重要的，这样开发人员可以集中精力创建业务逻辑，相应地缩短了开发时间。对于核心平台来讲，也就缩短了本地化开发的周期。

对异构环境的支持是基于 Java EE 体系的又一优秀特征。Java EE 能够开发部署在异构环境中的可移植程序。基于 Java EE 的应用程序不依赖任何特定操作系统、中间件、硬件。因此设计合理的基于 Java EE 的程序只需开发一次就可部署到各种平台。这在典型的异构企业计算环境中是十分关键的。Java EE 标准也允许客户订购与 Java EE 兼容的第三方的现成的组件，把它们部署到异构环境中，节省了由自己制定整个方案所需的费用。

可伸缩性也是一项重要的系统特征。平台应提供极佳的可伸缩性去满足那些在他们系统上进行业务处理的各种数量级别的数据。基于 Java EE 平台的应用程序可被部署到各种操作系统和平台上。例如可被部署到高端 UNIX 与大型机系统，这种系统单机可支持 64 至 256 个处理器，具备极强的处理能力；也可以部署在小型机上，在系统管理的参保人数增长到系统不能负担时，再增加服务器数量，进行集群；甚至，在一些性能要求较低的区域，还可以采用 PC Sever 作为应用服务器或数据库的载体。多级的应用配置不但满足了不同层面上对性能和成本的要求，而且也尽可能地减少不同平台之间迁移的费用。这在以往的系统规划中是不可想象的。

系统的稳定可用性是任何用户使用管理信息系统的终极目标，其信息数据的安全和稳定更是重中之重。一个服务器端平台必须能全天候运转以满足公司客户、合作伙伴的需要。若是意外停机，那会有灾难性后果。Java EE 体系的稳定性是基于 Java 的健壮性和虚拟机（Virtual Machine，VM）实现的一致性基础上的。Java EE 部署到可靠的操作环境中，它们支持长期的可用性。一些 Java EE 部署在 Windows 环境中，客户也可选择健壮性能更好的操作系统如 Sun Solaris、IBM OS/390。这是实时性很强的业务系统最理想的选择。

1.6　Java EE 主要技术

Java EE 平台由一整套服务（Services）、应用程序接口（API）和协议构成，它对开发基于 Web 的多层应用程序提供了强大的功能支持，下面对 Java EE 中的核心技术规范进行简单的描述。

1.6.1　JDBC（Java Database Connectivity）

JDBC API 为访问不同的数据库提供了统一的途径，像 ODBC 一样，JDBC 对开发者屏蔽了一些细节问题，另外，JDBC 对数据库的访问也具有平台无关性。

JDBC API 以一个统一的方式访问各种数据库，其接口包含在 java.sql 和 javax.sql 这两个包中。与 ODBC 类似，JDBC 将开发者和私有数据库隔离开来。由于建立在 Java 上，JDBC 还可以提供与平台无关的数据库访问。JDBC 定义了 4 种不同的驱动，具体来说，包括：

- JDBC-ODBC 桥

在 JDBC 刚产生时，JDBC-ODBC 桥是非常有用的。通过它，开发者可以使用 JDBC 来访问一个 ODBC 数据源。缺点是，它需要在客户机器上安装一个 ODBC 驱动，该机器通常运行的是微软 Windows 系统。使用这一类的驱动器，你就会失去 JDBC 平台无关的好处。此外，ODBC 驱动器需要客户端的管理。

- JDBC-native 驱动桥

JDBC-native 驱动桥提供了一个构建在本地数据库驱动上的 JDBC 接口而没有使用 ODBC。JDBC 驱动将标准的 JDBC 调用转化为对数据库 API 的本地调用，使用这一类的驱动也会失去 JDBC 平台无关性的好处，并且需要安装客户端。

- JDBC-network 桥

JDBC-network 桥不需要客户端的数据库驱动。它使用网络上的中间服务器来访问一个数据库，这种应用使得很多技术的实现有了可能，这些技术包括负载均衡、连接缓冲池和数据缓存等。由于第 3 种类型往往只需要更少的下载时间，具有平台独立性，而且不需要在客户端安装并取得控制权,所以很适合于 Internet 上的应用。

● 纯 Java 驱动

这一类驱动使用纯 Java 数据库驱动来提供直接的数据库访问,由于这类驱动运行在客户端,并且直接访问数据库,因此在这个模式下要使用一个两层的体系。要在一个 n 层的体系中使用该类型的驱动,可以通过一个包含有数据访问代码的 EJB,并且让该 EJB 为它的客户提供一个与数据库无关的服务。

企业级应用的另一个常见的数据库特性是事务处理。事务是一组申明(statement),它们必须作为同一个 statement 来处理以保证数据完整性。缺省情况下,JDBC 使用自动提交(auto-commit)事务模式,这可以通过使用 Connection 类的 setAutoCommit() 方法来实现,一个自动提交意味着在每一个数据库读写操作之后任何其他应用程序显示数据时都会看到更新了的数据。然而,如果你的应用程序执行两部分相互依赖的数据库访问操作,你可能会想要用 JTA API 去确定整个事务,这个事务将包含两个操作的开始、回滚和提交。

有关 JDBC 的更多信息,请参考 JDBC 指南:http://java.sun.com/products/jdbc/。

1.6.2　JNDI(Java Name and Directory Interface)

JNDI API 用于执行名字和目录服务,其接口包含在 javax.naming 和它的子包中。它为应用程序提供标准的目录操作的方法,例如获得对象的关联属性、根据它们的属性搜寻对象等。使用 JNDI,一个 Java EE 应用程序可以存储和动态获取任何类型的命名 Java 对象。

因为 JNDI 不依赖于任何特定的执行,应用程序可以使用 JNDI 访问各种命名目录服务,包括现有的 LDAP、NDS、DNS、NIS、COS 命名和 RMI 注册等服务。这使得 Java EE 应用程序可以和传统的应用程序和系统共存。

JNDI 分为两部分:应用程序编程接口(API)和服务供应商接口(SPI),前者允许 Java 应用程序访问各种命名和目录服务,后者则是通过设计来供任意一种服务的供应商(也包括目录服务供应商)使用。这使得各种各样的目录服务和命名服务能够透明地插入到使用 JNDI API 的 Java 应用程序中。

有关 JNDI 的更多信息,请参考 JNDI 指南:http://java.sun.com/products/jndi/。

1.6.3　EJB(Enterprise JavaBean)

Java EE 技术赢得广泛重视的主要原因之一就是 EJB(Enterprise JavaBean),要注意的是 EJB 从技术上而言不是一种产品,而是一个技术规范。它提供了一个框架来开发和实施分布式商务逻辑,由此简化了具有可伸缩性和高度复杂的企业级应用的开发。此外,它还定义了 EJB 组件在何时、如何与它们的容器进行交互作用,容器负责提供公用的服务,例如:目录服务、事务管理、安全性、资源缓冲池以及容错性。有关 EJB 框架的 API 定义在 javax.ejb 及其子包 javax.ejb.spi 中。

为了弄清楚 Enterprise JavaBean 的概念,可以先将它与 Java 常用的 JavaBean 做比较。JavaBean 是 Java 的组件模型,在 JavaBean 规范中定义了事件和属性等特征。Enterprise JavaBean 也定义了一个 Java 组件模型,但 Enterprise JavaBean 组件模型和 JavaBean 组件模型是不同的。JavaBean 的重点是允许开发者在开发工具中可视化地操纵组件,它解释了组件间的事件登记、传递、识别和属性使用、定制和持久化的应用编程接口。Enterprise JavaBean 的侧重点则是详细地定义了一个可以移植的 Java 组件的服务框架模型。因此,其中并没提及事件——Enterprise Bean 通常不发送和接受事件;同样也没有提及属性——属性定制并不是在开发时进行,而是在运行时(实际上在部署时)通过一个部署描述符来描述。

1.6.4 RMI(Remote Method Invoke)

Remote Method Invocation(RMI,远程方法调用)为分布式计算提供了一种高级的通用解决方案。正如名字所显示的那样,RMI 即调用远程对象上的方法,它使用连续序列方式在客户端和服务器端传递数据,将面向对象编程模型扩展到了客户机/服务器系统,使开发者可以用本地对象调用的语法进行远程调用。RMI 在 java.rmi 及其子包中实现。

1.6.5 JSP(JavaServer Pages)

JavaServer Pages(JSP)提供的功能大多与 Java Servlet 类似,不过实现的方式不同。Servlet 全部由 Java 写成并且生成 HTML;而 JSP 通常是大多数 HTML 代码中嵌入少量的 Java 代码。

JSP 是对 HTML 的一种扩展,它可以通过一些特殊的标签向静态 HTML 页面中插入动态的信息。如可以利用 <% 和 %> 标签添加 Java 代码段,用 <%= 表达式 %> 将表达式的值写入页面,用 <jsp:Bean> 标签在某一范围内(request、session 或 context)引用 Java Bean。除此之外,JSP 的标准标签扩展机制还允许开发人员编写自己的标签和相应的实现方法。这样就可以将某些商业逻辑封装成 JSP 的标签,使 JSP 文件更加简单、易于实现。

当一个浏览器首次向服务器请求一个 JSP 文件时,这个 JSP 文件首先被 Web 应用服务器编译成 Servlet 并执行,然后将所产生的结果作为一个 HTML 文件传给浏览器。只要在 JSP 文件中加入一些控制,便可轻松实现对数据的动态显示。以后,如果再有对这个 JSP 文件的请求,且该文件没有做任何修改,它将不会再被编译,而是直接执行已编译好的 Servlet。

有关 JSP 的更多信息,请参考 JSP 指南:http://java.sun.com/products/jsp/。

1.6.6 Servlet

Java Servlet 实质上是一种小型的、与平台无关的 Java 类,它由容器管理并被编译成平台无关的字节代码,这些代码可以动态地加载到一个 Web 服务器上,并由该 Web 服务器运行。Servlet 通过一种由 Servlet 容器实现的请求—响应模型与 Web 客户机进行交互。这种请求-响应模型建立在超文本传输协议(HTTP)行为的基础之上。Java Servlet API 提供了一种通用机制,对于任何使用了基于请求—响应协议的服务器,这种机制可以扩展其功能。Java Servlet API 接口定义在 Java.Servlet 和 javax.servlet.http 包中。随着越来越多的 Web 服务器的支持,Servlet 逐渐取代了基于 Java 的 CGI 脚本语言。与 CGI 脚本语言相比,Servlets 在可伸缩性上提供了很好的改进:每一个 CGI 在开始的时候都要求开始一个新的进程,而 Servlets 在 Servlet 引擎中是以分离的线程来运行的。此外,相比于现在的竞争者 ASP 和 JavaScript,Servlet 的主要优点在于其平台无关性。

Java Servlet API 与其他 Java Enterprise API 不同之处在于它不是一个在现存的网络服务

和协议之上的 Java 层；相反，类似于 CGI 应用程序，它本身往往不是完整的应用程序。在处理接收来自 Web 浏览器上用户的信息请求时，CGI 只是整个处理过程中的一个中间步骤。

例如，CGI 应用程序的一种常见用途是访问数据库。将它用于这种任务时，CGI 程序提供一种方法，将用户的数据请求连接到能满足这种请求的企业数据库，并向用户发回请求结果作为响应。Java Servlet 与 CGI 程序相似，适合于充当连接前端 Web 请求与后端数据资源的分布式中间件。

有关 Java Servlet 的更多信息，请参考 Java Servlet 指南：http://Java.sun.com/products/Servlet/。

1.6.7 JMS（Java Message Service）

企业消息传递系统通常又称为面向消息的中间件（Message Oriented MiddleWare，MOM），它使用松散耦合的、非常灵活的方式来集成应用程序，在存储和转发的基础上支持应用程序间数据的异步传递；每个应用程序都只与作为中介的 MOM 通信。

JMS 就是用于和企业消息传递系统相互通信的应用程序接口。在 JMS 出现之前，每个 MOM 供应商都提供专有的 API 以供客户程序访问他们的产品；而 JMS 为 Java 客户程序通过 MOM 产品收发消息提供了一种标准可移植的方法——用 JMS 编写的程序能够在任何实现了 JMS 标准的 MOM 上运行。

1.6.8 JTA（Java Transaction API）

Java Transaction API 指定事务管理器与分布式事务中涉及的其他系统组件之间的各种高级接口，这些系统组件有应用程序、应用程序服务器和资源管理器等。JTA 功能允许事务由应用程序本身、应用程序服务器或一个外部事务管理器来管理。JTA 接口包含在 javax.transaction 和 javax.transaction.xa 这两个包中。

XA 接口定义了资源管理器和分布式事务环境中外部事务管理器之间的约定。外部事务管理器可以跨多个资源协调事务。XA 的 Java 映射包含在 Java Transaction API 中。

遗憾的是，使用 JTA 处理分布式事务仍然很繁琐且易出错，更多情况下它仅仅是服务器实现者所使用的低级 API，而不为企业级程序员所使用。

有关 Java 事务的更多信息，请参考 JTA 指南：http://www.Java.sun.com/products/jta/。

1.6.9 JavaMail

JavaMail 是用于存取邮件服务器的 API，它提供了一套邮件服务器的抽象类。不仅支持 SMTP 服务器，也支持 IMAP 服务器。

1.6.10 JAXP（Java API for XML Processing）

XML 是一种描述基于文本的数据的语言，使用 XML 使得数据可以被任何程序和工具读取和处理。通过程序和工具可以生成其他程序和工具可以读取和处理的 XML 文档。Java XML 处理 API（JAXP）支持使用 DOM、SAX 和 XSLT 对 XML 文档进行处理。JAXP 使得 Java 应用程序可以不依赖于特殊的 XML 处理执行、解析和转换 XML 文档。

例如，一个 Java EE 应用程序可以使用 XML 来生成报表，而不同的公司都可以获得这个报表并使用各自最适宜的方法来处理它。一个公司可能会通过程序将 XML 数据导入到 HTML 页面中以使其可以在网站中公布，另一个公司可能会通过工具导出 XML 数据以制定销售预算。

1.6.11　Java EE Connector Architecture

Java EE 工具提供商使用 Java EE 连接器体系结构，建立可以支持访问企业信息系统的资源适配器。一个资源适配器就是一个使 Java EE 应用程序组件可以访问底层的资源管理器并与其实现交互的软件组件。因为一个资源适配器是与其特定的资源管理器相对应的，典型的情况是不同的数据库或企业信息系统会自有其不同的资源适配器。

1.7　安装 WebLogic 服务器

作为 BEA WebLogic Enterprise Platform 产品中最重要的一部分，WebLogic 服务器为开发和部署 Java EE 的应用程序提供了强有力的支持。BEA WebLogic 服务器提供所有核心应用服务器应该提供的功能和服务，例如：

- 负载平衡（Load balancing）
- 容错（Fault tolerance）
- Web 服务（Web Services）
- 与大型机结合（Legacy integration）
- 事务管理（Transaction management）
- 安全（Security）
- 多线程（Multi-threading）
- 持久性（Persistence）
- 数据库连接（Database connectivity）
- 资源池化（Resource pooling）

1.7.1　安装前的准备

- 下载软件

你可以从 BEA 网站上（http://commerce.bea.com）下载 WebLogic Server 的评估版本，支持的操作系统有 Microsoft Windows XP/7/8、Linux、Unix 和 Solaris 等，也可以在此网址下载 WebLogic 服务器的旧版本。下载的时候需要免费注册成为有效的用户才可以下载，有两种下载安装方法：

Package installer—下载一个 EXE 文件，这个 EXE 是一个打包的文件，包括了所有的 WebLogic 服务器软件的各个部分。例如：WebLogic 服务器和相关的例子，WebLogic Workshop 和相关的例子，WebLogic JRockit。这个 EXE 文件大概有 230 MB。

Net installer—先下载一个 EXE 文件，然后运行该程序，该程序可以让你有选择地安装相关的部件。然后在线下载你选择安装的部件。这个 EXE 文件大概有 22 MB。

推荐大家使用 Package installer 下载方式。

● 系统需求

操作系统：可以在 Windows NT、Windows 7、Windows 8、Unix、Linux 等操作系统安装。

硬盘空间：大概需要 650 MB。

内存：最少 256 MB，推荐使用 512 MB 或更多的内存。

管理员权限：如果你想在 Windows 操作系统下，把 WebLogic 服务器以服务形式运行，那么你需要有管理员的权限进行这项设置。

许可协议（License）：WebLogic Server 的运行不能没有相应的许可协议。当你安装 WebLogic Server 后，安装程序自动创建了一个可以评估使用的许可协议，该许可协议可以最多有 5 个并发的客户连接、15 个数据库连接，一年的使用有效期。

1.7.2 配置 WebLogic 服务器

从开始菜单中运行 Configuration Wizard，配置一个 WebLogic 服务器的域（domain）。域在 WebLogic 当中是一个基本的管理单元，一个域可以包含一个或多个运行着的 WebLogic 服务器，而这些 WebLogic 服务器和它们相关的资源都由单一的管理服务器（Administration Server）进行管理。一个最小的域可以只包含一个 WebLogic 服务器，这个 WebLogic 服务器既是管理服务器又是被管理的服务器（Managed Server），这种域通常在开发阶段被采用，但是在实际的产品运行阶段一个域通常会有单一的管理服务器和若干个被管理的服务器。

在弹出的窗口中，选择创建一个新的设置。

● 修改环境变量

右击桌面上"我的电脑"的图标，选中"属性"，在弹出的窗口中，选中"高级"选项卡，在"高级"选项卡中点"环境变量"，先把用户变量中 TEMP 和 TMP 的参数修改成和系统变量 TEMP 和 TMP 一样的数值，例如：C:\WINNT\TEMP。

在系统变量中，点"新建"按钮，增加一个 WL_HOME 的参数，数值为你安装的 BEA Weblogic 服务器的目录，例如：c:\bea\weblogic81\server。

修改系统变量中的 CLASSPATH 参数，加上 webloigic.jar 文件的路径。如果电脑中的系统变量中没有 CLASSPATH 参数，则添加一个 CLASSPATH 的参数，数值为 dt.jar、tools.jar 和 webloigic.jar 文件的路径：c:\bea\jdk16\lib\dt.jar；c:\jdk16\lib\tools.jar；c:\%WL_HOME% \lib\weblogic.jar。

修改系统变量中的 path 参数，加上 jdk 中 bin 目录的路径。例如：c:\bea\jdk16\bin。

● 验证环境变量的修改

选中"开始"菜单中的运行，在弹出的"运行"菜单中输入 cmd，在弹出的 DOS 窗口中运行 javac，在弹出的 DOS 窗口中运行 java weblogic.ejbc –version。

1.8 部署 Web 应用程序

1.8.1 部署

在 Java EE 部署体系中，表示 Java EE 平台产品所需要的动态部署配置信息的组件采用了 JavaBean 结构，因为这种结构既适于表示简单组件，又适于表示复杂组件，同时它还有很强的平台无关性。这些 Bean 使开发简单的属性页、编辑器和复杂的定制向导（它可以引导部署者完成应用程序部署配置各步骤）成为易事。

Java EE 部署 API 包括如下内容：

Java EE 平台产品必须实现的一套最小工作集。所有的 Java EE 平台产品提供商都必须向工具提供商提供这套工作集的实现，它使得可移植应用程序可部署到不同的 Java EE 平台产品上。

部署工具所必须实现的一套最小工作集。所有的 Java EE 工具提供商都必须提供这套工作集的实现，以与不同的 Java EE 平台产品交互。

这套 API 描述了三个部署步骤中的两步：安装与配置，第三步（运行）留给了 Java EE 平台产品提供商。这些提供商可以在其自己的部署工具中扩展上述最小工作集以与其他厂商竞争，这些扩展可能对其他厂商的部署工具不可用。

关于 Java EE 部署的更多信息，请参考 Java EE 部署指南：http://Java.sun.com/j2ee/tools/deployment/。

1.8.2 Web 应用程序

http 以及 Web Services 的打包和部署遵照 Sun 公司的 Web 应用程序标准。该标准定义 Web 应用程序作为各种基于 Web 应用的组件组合在一起的标准。这些组件可以是 JSP 页面，Servlets 以及诸如 HTML 页面与图像文件等静态资源。Web 应用可以访问外部资源，例如 EJB 与 JSP 标记库。一个服务器中可以存放任意多个 Web 应用程序。当你从 Web 应用程序中请求资源时，一般都要在请求的 URI 中包含该应用程序的名字。

一个 Web 包文件包含了所有构成该 Web 应用程序的文件。一个 .war 文件作为一个单元被部署到 WebLogic 服务器中。WebLogic 服务器中的一个 .war 文件可以包含如下内容：
- 至少有一个 Servlet 或 JSP 页面，以及它们的帮助类（helper classes）。
- web.xml 部署描述文件，一个符合 Java EE 规范，描述 .war 文件内容的标准 XML 文件。
- weblogic.xml 部署描述文件，一个为 Web 应用指定 WebLogic 服务器所要求的 XML 文件。
- .war 文件还可能包含 HTML/XML 页面及其支持文件（例如图片和多媒体文件）。

1.9 小结

- ✓ 对 Java EE 的整体架构做一个全局性的介绍。
- ✓ 介绍了应用服务器的概念。
- ✓ WebLogic 服务器的安装、配置和部署 Web 应用程序的基本步骤。

1.10 英语角

client	客户端
server	服务器
container	容器
Enterprise JavaBean	企业级 JavaBean

1.11 作业

1. 安装和配置 WebLogic 服务器。
2. 部署 Web 应用。

1.12 思考题

Java EE 的优势主要体现在哪些方面？

1.13 学员回顾内容

1. 什么是多层结构？
2. 什么是 Java EE 容器？

第2章 使用Ajax改进用户体验

学习目标

- ✧ 了解XMLHttpRequest对象的作用。
- ✧ 理解XMLHttpRequest对象的属性和方法。
- ✧ 掌握XMLHttpRequest对象的应用。

课前准备

- ✧ 在互联网上搜索有关Ajax的表单验证的资料。
- ✧ 知道什么是动态加载列表框。
- ✧ 知道什么是自动完成功能。

本章简介

在第1章中我们介绍了Java EE的基本概念，包括Client/Server结构，多层结构，Java EE容器、开发角色等，Java EE核心的API和Java EE中不同的开发角色。在学习了这些之后，我们会发现Ajax在应用中的地位。本章通过用户注册案例和信息搜索案例来了解Ajax的作用、用途及掌握XMLHttpRequest对象的应用。

2.1 Web 3.0

2.1.1 互联网的前世今生

到今天，互联网已经渗透到了我们生产、生活的方方面面。互联网的发展经历了以下几个阶段：

● 黑暗年代：数据和信息不通过电子途径传播和共享的年代。在黑暗年代里，纸、笔、打字机、大型的印刷机器、广播、电视是主要的信息载体，而且当时人们也没有意识到信息究竟有多大的价值。

● 静态网页时代：早期的互联网仅提供给人们静态页面进行浏览。这已经让人们欣喜若狂了。人们从没想到信息可以传播那么广，那么远。互联网迅速发展，一夜之间占领了这个星球。

● 动态网页时代：技术的发展从未止步，动态网页技术的发展使人们可以在互联网上发布信息，数据库技术的发展使互联网变得越来越实用。Web 开始渗透到了生产领域，传统的 C/S 模式被 B/S 模式取代，Web 开始和企业级技术紧密结合。

● Web 2.0 时代：如果说互联网的出现是一次革命，那 Web 2.0 轰轰烈烈地登场就是另外一次革命。在 Web 2.0 时代里，我们追求更佳的用户体验和更高的参与程度。

● Web 3.0 时代：Web 2.0 是具有革命性意义的。人们在创造劳动的同时将获得更多的荣誉、认同，包括财富和地位。正是因为更多的人参与到了有价值的创造劳动，那么"要求互联网价值的重新分配"将是一种必然趋势，因而必然促成新一代互联网的产生，这就是 Web 3.0。

2.1.2 Web 3.0 概述

Web 3.0 有别于以往，主要体现在以下几个方面。

● 主动性

Web 3.0 最明显的特征就是主动性，即强调网站对用户需求的主动提取，并加以分析处理，然后给出用户所需要的资源。这点类似于新浪网推出的智能交互式搜索引擎 iask，用户可以将自己的需求通过问题的形式提出，然后借助 iask 的海量知识库和用户回答两种形式给出答案。但是，搜索引擎仅仅是解决了一个问题而已，它无法直接解决用户生活或者工作中所遇到的具体的困难，这就需要更为专业的服务型站点来提供一站式服务。这样，用户不仅通过互联网获取答案，还能直接接受服务以便解决更为复杂的需求。

目前网上已经出现具有明显 Web 3.0 特征的网站了，如：如易网，它有 15 个大门类，175 个小门类，让用户能实现对目标对象的直接定位，可实现目标客户准确定位、客户服务、客户管理、营销等一系列经营环节。

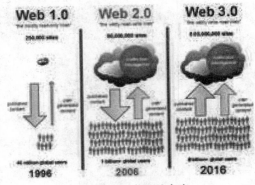

图 2-1 用户贡献内容

● 数字最大化

Web 3.0 另外一个主要特征为数字最大化。无论是商品还是服务的交易，对商家来说都会涉及时间效率和空间成本的问题。而通过互联网按照一定规则将商品和服务数字化，同时解决时空上的矛盾。

目前云网在商品数字化方面探索相当成功，云网是一家主要经营游戏点卡、电信卡等数字商品的专业电子商务公司，其看准了点卡容易数字化的特点而取得了很大成功。而随着信息化程度的不断加深，会有更多的商品以数字化的方式出现，比如电子票、电子书、电子书包等。

服务类的数字化方面,如易网也同样做了有效的探索与尝试。对部分服务领域如教育培训、咨询等领域将提供视频模块,实现线上服务,从而达到服务数字化的目的,使得服务的整个过程完全在网络环境下进行。

● 多维化

近几年,随着网络资源的丰富,多媒体技术在互联网上的应用越来越广泛。但主要集中在一些娱乐领域,比如在线视频电影,网络游戏等。实际上 Web 3.0 时代将会出现信息多维化的趋势。比如目前拍卖网站的商品描述主要通过图片和文字的形式表现,如果为部分高端商品提供视频描述或者三维动画,不仅为消费者提供更全面的信息,而且也会提升该商品的品质,从而更有利于商品的出售。

2.2 Ajax

2.2.1 为什么用 Ajax

Ajax 技术使我们可以通过 JavaScript 发送请求到服务器,并获得返回结果。这就让我们可以在必要的时候只更新页面的一小部分,而不用整个页面都刷新,这称为"无刷新"技术。如图 2-2 所示,搜狐首页上的登录功能就使用了 Ajax 技术。输入登录信息单击"登录"按钮后,只是刷新登录区域的内容。首页上信息很多,局部刷新就避免了重复加载,浪费网络资源。这是无刷新技术的一个优势。

图 2-2 使用 Ajax 刷新局部页面

再看土豆网的例子。在观看视频的时候,我们可以在页面上单击其他按钮执行操作,由于只是局部刷新,页面不会刷新,视频继续播放,不会受影响,如图 2-3 所示。这体现了无刷新技术的第二个优势:提供连续的用户体验,不因页面刷新而中断。

图 2-3　使用 Ajax 无刷新观看视频

最后看 Google 地图的例子。由于采用了无刷新技术,我们可以实现一些以前 B/S 程序很难做到的事情,比如图 2-4 中 Google 地图提供的拖动、放大、缩小等类似桌面程序的用户体验。

图 2-4　类似桌面程序的用户体验

2.2.2　什么是 Ajax

我们已经知道使用 Ajax 技术可以通过 JavaScript 发送请求到服务器,根据返回结果只更新局部页面,那什么是 Ajax 呢?

如图 2-5 所示,Ajax 是几个英文单词首字母的缩写:Asynchronous JavaScript And Xml。Ajax 并不是一种全新的技术,而是整合了几种现有的技术:JavaScript、XML 和 CSS。主要是 JavaScript。我们通过 JavaScript 的 XMLHttpRequest 对象完成发送请求到服务器并获得返回结果,然后使用 JavaScript 更新局部的网页。Asynchronous(异步)指的是 JavaScript 脚本发送

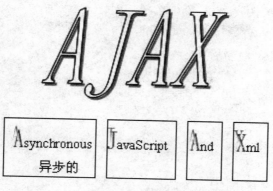

图 2-5 Ajax 组成

请求后并不是一直等着服务器响应,而是发送请求后继续做别的事,请求响应的处理是异步完成的。XML 一般用于请求数据和响应数据的封装,CSS 用于美化页面样式。

2.2.3 怎样使用 Ajax

我们用一个实际的案例解释 Ajax 的应用。在用户输入注册信息时,通常因为输入的用户名已被使用而一遍又一遍地重复注册,这对用户来说是不好的体验。如果当用户名输入框失去焦点时,就发送请求到服务器,判断用户是否存在,已经存在则弹出窗口提示:"用户名已经存在"。完成这个功能的过程中页面不刷新便可以解决这个难题了,如图 2-6 所示。

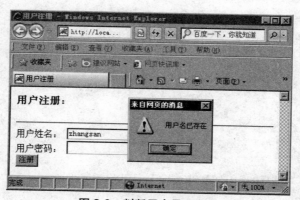

图 2-6 判断用户是否存在

使用输入框的 onblur 事件,实现输出框失去焦点时调用检查用户名是否存在的 JavaScript 方法,如示例代码 2-1 所示。

```
示例代码 2-1:index.jsp 页面源代码片段
<%@ page language="java" import="java.util.*" pageEncoding="UTF-8"%>
<!DOCTYPE HTML PUBLIC "-//W3C//DTD HTML 4.01 Transitional//EN">
<html>
    <head>
```

```
            <title> 用户注册 </title>
            <script type="text/javascript">
    function checkUserExists(oCtl){
            var uname = oCtl.value;
            if(uname ==""||name==null){
                alert(" 用户名不能为空 ");
                        oCtl.focus();
                    return false;
            }
            // 发送请求到服务器,判断用户名是否存在
        //...Ajax here...
            }
    </script>
    </head>

    <body>
        <h3>
                用户注册：
        </h3>
        <hr />
        用户姓名：
        <input type="text" name="username" value="" onblur="return checkUserExists(this)" />
        <span id="spanBlock"></span>
        <br>
        用户密码：
        <input type="password" name="userpass" value="" />
        <br>
        <input type="button" name="sub" value=" 注册 " />
        <br>
    </body>
</html>
```

　　我们现在还没有编写 Ajax 代码。在这之前,我们要先开发好服务器端的程序,在 UserServlet 中创建相应方法,如果 uname 参数指定的用户名存在则响应输出 true,否则响应输出 false。如示例代码 2-2 所示。

示例代码 2-2：UserServlet 源代码片段

```java
package com.xtgj.j2ee.chapter02.servlet;

import java.io.IOException;
import java.io.PrintWriter;
import java.sql.SQLException;

import javax.servlet.ServletException;
import javax.servlet.http.HttpServlet;
import javax.servlet.http.HttpServletRequest;
import javax.servlet.http.HttpServletResponse;

import com.xtgj.j2ee.chapter02.user.biz.UserManager;

public class UserServlet extends HttpServlet {
    public void doGet(HttpServletRequest request,HttpServletResponse response)
throws ServletException,IOException{
            response.setContentType("text/html;charset=utf-8");
            response.setHeader("Cache-Control", "no-cahce");
            PrintWriter out=response.getWriter();
            String text="";
            String uname = request.getParameter("uname");
            UserManager m = new UserManager();
            try {
                    if (m.check(uname)){
                            text="true";
                            out.print(text);
                    }
            }catch(SQLException e){
                    e.printStackTrace();
            }
            out.close();
    }
    public void doPost(HttpServletRequest request, HttpServletResponse response)
                    throws ServletException, IOException {
            doGet(request,response);
    }
}
```

处理用户名存在与否的验证逻辑类 UserManager 源代码如示例代码 2-3 所示。

示例代码 2-3：UserManager.java 源代码片段

```java
package com.xtgj.j2ee.chapter02.user.biz;
import java.sql.*;
import com.xtgj.j2ee.chapter02.db.DBUtil;
public class UserManager {
    Connection conn = null;
    Statement st = null;
    ResultSet rs = null;

    public UserManager() {
        this.conn = DBUtil.getConn();
    }

    public boolean check(String uname) throws SQLException {
        try {
            st = conn.createStatement();
            String sql = "select * from user_table where username='" + uname
                    + "'";
            rs = st.executeQuery(sql);
            return rs.next();
        } catch (Exception e) {
            e.printStackTrace();
        } finally {
            DBUtil.close(rs, st, conn);
        }
        return false;
    }
}
```

在 web.xml 中添加如下配置：

示例代码 2-4：web.xml 源代码片段

```xml
<servlet>
  <servlet-name>UserServlet</servlet-name>
  <servlet-class>com.xtgj.j2ee.chapter02.servlet.UserServlet</servlet-class>
</servlet>
```

```xml
<servlet-mapping>
    <servlet-name>UserServlet</servlet-name>
    <url-pattern>/UserServlet</url-pattern>
</servlet-mapping>
```

现在我们使用 JavaScript 的 XMLHttpRequest 对象发送请求。创建 XMLHttpRequest 对象的代码如示例代码 2-5 所示。

示例代码 2-5：创建 XMLHttpRequset 对象的源代码片段

```javascript
function createXMLHttpRequset() {
    if (window.XMLHttpRequest) {
        return new XMLHttpRequest();
    } else if (window.ActiveXObject) {
        try {
            return new ActiveXObject("Msxml2.XMLHTTP");
        } catch (e) {
            try {
                return new ActiveXObject("Microsoft.XMLHTTP");
            } catch (e) {
            }
        }
    }
}
```

使用 XMLHttpRequest 对象分 4 步完成，本例中，我们定义 ajax.js 文件封装这些操作。如示例代码 2-6 所示。

示例代码 2-6：ajax.js 文件源代码

```javascript
var xmlHttpRequest;
function checkUserExists(oCtl) {
    var uname = oCtl.value;
    if (uname == "" || uname == null) {
        alert(" 用户名不能为空 ");
        oCtl.focus();
        return false;
    }
    doRequest(uname);
}
function createXMLHttpRequset() {
```

```
        if (window.XMLHttpRequest) {
            return new XMLHttpRequest();
        } else if (window.ActiveXObject) {
            try {
                return new ActiveXObject("Msxml2.XMLHTTP");
            } catch (e) {
                try {
                    return new ActiveXObject("Microsoft.XMLHTTP");
                } catch (e) {
                }
            }
        }
    }
    function doRequest(uname) {
        // 请求字符串
        var url = "UserServlet?uname=" + uname;
        // 1. 创建 xmlHttpRequest 组件
        xmlHttpRequest = createXMLHttpRequset();
        // 2. 设置回调函数
        xmlHttpRequest.onreadystatechange = haoLeJiaoWo;
        // 3. 初始化 xmlHttpRequest 组件
        xmlHttpRequest.open("GET", url, true);
        // 4. 发送请求
        xmlHttpRequest.send(null);
    }
    function haoLeJiaoWo() {
        if (xmlHttpRequest.readyState == 4) {
            if (xmlHttpRequest.status == 200) {
                var b = xmlHttpRequest.responseText;
                if (b == "true") {
                    alert(" 用户名已存在 ");
                }else{
                    document.getElementById("spanBlock").innerHTML = " 用户名可用 ";
                }
            } else {
                alert(" 您请求的页面有异常 ");
            }
```

```
    }
}
```

解释一下第 2 步的代码"XMLHttpRequest.onreadystatechange=haoLeJiaoWo;": onreadystatechange 是 XMLHttpRequest 对象的一个事件,haoLeJiaoWo 是一个 JavaScript 方法名。这句代码的含义是:在 XMLHttpRequest 对象的状态发生改变时,调用 haoLeJiaoWo 这个方法。

XMLHttpRequest 对象的 open 方法用于初始化 XMLHttpRequest 对象,有 3 个参数:第一个参数可以是"POST"或"GET",用于指定请求发送的方式;第二个参数用于指定请求的 url;第三个参数表明请求是同步的还是异步的,true 表示是异步的,意思是在调用 send 方法发送请求后不用等请求响应继续执行后面的 JavaScript 语句。

现在我们看看 haoLeJiaoWo 这个函数。每次 XMLHttpRequest 对象状态改变时就会调用这个方法。XMLHttpRequest 对象有 5 种状态:0- 未初始化;1- 已初始化;2- 发送请求;3- 开始接收结果;4- 接收结果完毕。所以我们只处理状态 (readyState) 为 4 的情况。XMLHttpRequest.status 表示服务器响应的状态码。200 表示正常,一般常见的还有 404,表示 Not found(没有找到资源),500 表示服务器端出错。我们只处理正常响应的情况,即 XMLHttpRequest.status 为 200 的情况。当 XMLHttpRequest 的 readyState 为 4 且 status 为 200 时,我们就可以从 XMLHttpRequest.responseText 中得到服务器响应的字符串了,而后再做进一步的处理。

在 index.jsp 中修改 <script> 节点如下:

```
<head>
        <base href="<%=basePath%>">
        <title> 用户注册 </title>
        <script type="text/javascript" src="ajax.js"></script>
</head>
```

在这个案例中,数据库连接工厂类 DBUtil 源代码如示例代码 2-7 所示。

示例代码 2-7:数据库连接工厂类 DBUtil.java 源代码片段

```java
package com.xtgj.j2ee.chapter02.db;

import java.sql.Connection;
import java.sql.DriverManager;
import java.sql.ResultSet;
import java.sql.SQLException;
import java.sql.Statement;
import java.util.ResourceBundle;
public class DBUtil {

    private static String driver = "";
```

```java
private static String url = "";
private static String username = "";
private static String userpass = "";

private static DBUtil db = new DBUtil();

private DBUtil() {
    this.init();
    try {
        Class.forName(driver);
    } catch (ClassNotFoundException e) {
        pfe(e);
    }
}

/**
 * 用于获取数据库连接对象
 * @return
 */
public static Connection getConn() {
    Connection conn = null;
    try {
        conn = DriverManager.getConnection(url, username, userpass);
    } catch (SQLException e) {
        pfe(e);
    }
    return conn;
}

/**
 * 用于关闭数据库对象
 * @param rs
 * @param stmt
 * @param conn
 */
public static void close(ResultSet rs, Statement stmt, Connection conn) {
    try {
```

```
                    if (rs != null)    rs.close();
                    if (stmt != null)  stmt.close();
                    if (conn != null)  conn.close();
            } catch (SQLException e) {
                    pfe(e);
            }
    }

    /**
     * 这里要求 db.properties 文件必须在 src 根目录下
     */
    private void init() {
            ResourceBundle rb = ResourceBundle.getBundle("db");
            driver = rb.getString("driverClassName");
            url = rb.getString("url");
            username = rb.getString("userName");
            userpass = rb.getString("password");
    }

    private static void pfe(Exception e) {
            System.out.println("===" + e.getMessage() + "===");
    }
}
```

域类 User.java 源代码如示例代码 2-8 所示。

示例代码 2-8：实体类 User.java 源代码

```
package com.xtgj.j2ee.chapter02.entity;

public class User {
        private int id;
        private String username;
        private String password;
    public int getId() {
            return id;
    }
        public void setId(int id) {
                this.id = id;
```

```java
}
public String getUsername() {
        return username;
}
public void setUsername(String username) {
        this.username = username;
}
public String getPassword() {
        return password;
}
public void setPassword(String password) {
        this.password = password;
}

public User()
{
}
public User(int id, String username, String password) {
        this.id = id;
        this.username = username;
        this.password = password;
}
```

本例是一个与数据库交互的应用,我们使用 SqlServer 数据库管理系统,创建数据库 test,并在 test 数据库中创建数据库表 user_table,创建脚本如下所示:

```sql
create table user_table(
id int identity primary key,
username varchar(20) not null,
password varchar(20) not null
);
```

2.3 Ajax 技术结构

为了能更清楚地了解同步交互方式和异步交互方式的根本区别,下面对传统的处理方式和 Ajax 提供的处理方式进行一个比较。

2.3.1 传统 Web 应用解决方案

传统的 Web 应用是同步交互的方式,这种同步交互方式的处理过程如图 2-7 所示。

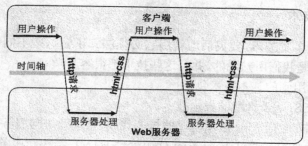

图 2-7 传统的处理方式

当用户向 Web 服务器提交了一个处理请求后,等待服务器的响应,服务器端将在接收到该请求后,按照预先编写的程序中的业务逻辑进行处理。显然,这样的处理方式一定会给用户一种不连贯的体验,因为当服务器在处理请求的时候,用户多数时间只能处于等待状态,页面中显示的内容也只能是一片空白。

2.3.2 Ajax 解决方案

与传统的 Web 应用不同,Ajax 采用的是一种异步交互的处理方式,这种异步交互方式的处理过程如图 2-8 所示。

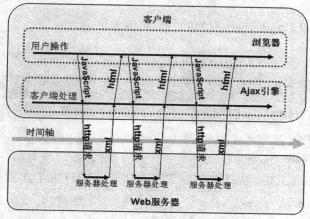

图 2-8 Ajax 的处理方式

Ajax 相当于在浏览器客户端与服务器之间架设了一个桥梁，在它的帮助下，可以消除网络交互过程中的"处理－等待－处理－等待"这样的缺陷。在处理过程中，Web 服务器响应是标准的且易于解析的 XML 格式的数据，该 XML 数据被传递给 Ajax，然后再转换成 HTML 页面的格式，辅助 CSS 进行显示。

Ajax 是使用 XMLHttpRequest 对象发送请求并获得服务器端的响应，同时 Ajax 可以在不重新载入整个页面的情况下用 JavaScript 操作 DOM 以实现页面更新。因此在读取数据的过程中，用户所面对的不是白屏，而是原来的页面内容。这种更新是瞬间的，用户几乎感觉不到，对用户来讲这是一种连贯的感受。

借助 Ajax，可以把一些原来由服务器负担的工作转移到客户端来完成，利用客户的闲置能力来进行处理，这样也可以有效地减轻服务器和带宽的负担，节约空间和宽带租用成本。

2.3.3 Ajax 相关技术

Ajax 实际上涉及多种技术，要灵活地运用它必须深入了解这些不同的技术。令人愉悦的是其中大部分技术都是我们所熟悉的，并且这些技术都很容易学习。

Ajax 应用程序所用到的基本技术包括：
- XHTML 和 CSS：基于标准的表示技术。
- JavaScript 脚本：运行 Ajax 应用程序的核心代码，帮助改进与服务器应用程序的通信。
- 文档对象模型 DOM：通过 JavaScript 代码处理 HTML 结构和服务器返回的 XML。
- XML 和 XSTL：进行浏览器和服务器两端的数据信息交换与处理。
- XMLHttpRequest：进行浏览器和服务两端的异步数据读取。

其中 XMLHttpRequest 是 Ajax 技术的核心对象，而 JavaScript 则是将其他几种技术结合在一起。为了对 Ajax 有一个概要的了解，下面介绍这几种技术。

2.3.4 XMLHttpRequest 对象的方法和属性

由于 JavaScript 具有动态类型特性，而且 XMLHttpRequest 在不同浏览器上的实现是兼容的，所以可以用同样的方式访问 XMLHttpRequest 实例的属性和方法，而不论这个实例创建的方法是什么。这都大大简化了开发过程。而且在 JavaScript 中也不必编写特定于浏览器的逻辑。

表 2-1 显示了 XMLHttpRequest 对象的一些典型方法。

表 2-1 XMLHttpRequest 对象的一些典型方法

方法	描述
Open("method", "url")	建立对服务器的调用
send (content)	向服务器发送请求
abort ( )	停止当前请求
getResponseHeader ("header")	返回指定响应首部的串值
setRequestHeader("header", "value")	设置指定首部的值，在设置前必须调用 open( )

这里我们对 open( ) 和 send( ) 方法加以详细说明。

> void open(String method,String url,boolean async,String username,String password);

这个方法会建立对服务器的调用。它有两个必要的参数,包括请求的特定方式(GET、POST 或 PUT),还要提供所请求资源的 URL。另外还有 3 个可选参数,包括传递一个 boolean 值,指示这个请求是异步的还是同步的,默认值为 true,表示请求本质上是异步的。如果将这个参数设置为 false,处理就会等待,直到从服务器返回响应为止。从某种程度上讲,这与我们使用 XMLHttpRequest 对象的初衷不太相符。不过,在某些情况下这个参数设置为 false 也是有用的。最后两个参数不说自明,允许你指定一个特定的用户名和密码。

> void send(content);

这个方法具体向服务器发送请求。按照顺序,open( ) 方法调用完后要调用这个方法。如果请求是异步的,这个方法就会立即返回,否则就是等待,直到接收到响应为止。send( ) 方法有一个参数,如果请求是以 POST 方式发出,则该参数可以是任何想传给服务器的请求内容,如果是以 GET 方式发出,由于请求内容附在 URL 之后,所以该参数通常置为 null。

除了上面这些标准方法,XMLHttpRequest 对象还提供了许多属性,如表 2-2 所示。在处理 XMLHttpRequest 对象时可以大量使用这些属性。

表 2-2　XMLHttpRequest 对象的一些典型属性

属性	描述
readyState	请求的状态。有 5 个可选的值: 0= 未初始化,1= 正在加载,2= 已加载,3= 交互中,4= 完成
status	服务器的 http 状态码: 200 对应 OK,404 对应 Not Found 等
statusText	http 状态码的相应文本: OK 或 Not Found 等
onreadystatechange	每个请求状态改变时都会触发这个事件处理器,通常会调用一个 JS 函数
responseText	服务器的响应,表示为一个字符串
responseXML	服务器的响应,表示为 XML,可以解析为一个 DOM 对象

2.3.5　发送请求

前面,我们了解了 XMLHttpRequest 对象的创建及其相关的方法和属性,现在我们来看如何向服务器发送请求,以及如何处理服务器的响应。请观察下列代码:

```
function createQueryString(){
    var queryStr = "userName=zhangsan";
    return queryStr;
}
function doRequest(){
    createXMLHttpRequset();
    var url="login?"+createQueryString();
    xmlHttp.open("GET",url,true);
    xmlHttp.onreadystatechange=handleResponse;
    xmlHttp.send(null);
}
```

我们来具体分析一下发送请求的各个步骤：

首先，为得到 XMLHttpRequest 对象实例的一个引用，可以创建一个新的实例。这里，我们只需调用这个函数即可，即 createXMLHttpRequest( )；。

接着，告诉 XMLHttpRequest 对象，哪个函数会处理 XMLHttpRequest 对象状态的改变，为此要把对象的 onreadystatechange 属性设置为指向 JavaScript 函数的指针，即 XMLHttp.onreadystatechange= handleResponse;。

然后，指定请求的属性。XMLHttpRequest 对象的 open( ) 方法会指定发出的请求。open( ) 方法通常取 3 个参数：一个是指示所用的方法（通常是 GET 或 POST）的串，一个是表示目标资源 URL 的串，一个是 boolean 值，指示请求是否是异步的。上述代码片段采用 GET 请求方式，即 XMLHttp.open("GET",url,true);。

POST 方式不同于 GET，需要设置 XMLHttpRequest 对象的 Content-Type 首部，如下列代码所示：

```
function doRequest( ){
    createXMLHttpRequset( );
    var url="login";
    xmlHttp.open("POST",url,true);
    xmlHttp.onreadystatechange=handleResponse;
    xmlHttp.setRequestHeader("Content-Type"
        ,"application/x-www-form-urlencoded");
    xmlHttp.send(createQueryString());
}
```

最后，将请求发送给服务器。XMLHttpRequest 对象的 send( ) 方法把请求发送给指定的目标资源。send( ) 方法接受一个参数，通常是一个串或一个 DOM 对象。这个参数作为请求体的一部分发送到目标 URL。当向 send( ) 方法提供参数时，要确保 open( ) 中指定的方法是 POST。如果没有数据作为请求体的一部分被发送，则使用 null。

对于 XMLHttpRequest 对象，onreadystatechange 属性存储了回调函数的指针，当 XMLHttpRequest 对象的内部状态发生变化时，就会调用这个回调函数。当进行了异步调用，请求就会发出，脚本立即继续处理（不必等待请求结束）。一旦发出了请求，对象的 readyState 属性会经过几个变化。尽管针对任何状态都可以做一些处理，不过我们最感兴趣的状态可能是服务器响应结束时的状态。如下代码所示：

```
function handleResponse( ){
    if(xmlHttp.readyState==4){// 服务器响应结束,信息已返回
        if(xmlHttp.status==200){// 确认返回页面有没有错误
            // 可以开始处理
        }else{
            // 页面有异常
        }
    }else{
        // 信息还没返回,等待
    }
}
```

通过设置回调函数，就可以有效地告诉 XMLHttpRequest 对象，"只要响应到来，就调用这个函数来处理响应"。

2.3.6 响应数据

我们在介绍 XMLHttpRequest 对象的典型属性时提到了服务器响应信息。服务器响应信息有两种格式，一种是 XML 格式，一种是简单文本格式。如果将服务器响应作为简单文本来访问，可以配合 HTML DOM 对象的 innerHTML 属性来共同完成信息的动态显示。例如本章第二小节介绍的用户名检测的案例，响应数据的格式就为简单文本，当用户名可用时，利用 <span> 显示可用的提示，得到如图 2-9 的显示结果。

图 2-9 用户名可用的效果图

服务器不一定都要按简单文本格式发送响应。当 Content-Type 响应首部正确地设置为

text/xml(如果是简单文本,Content-Type 响应首部则是 text/plain),将响应作为 XML 数据库发送完全是可以的(复杂的数据结构适合以 XML 格式发送)。此时使用 XMLHttpRequest 对象的 responseXML 属性来获取响应数据。

2.4 信息搜索示例

这一节,我们将综合运用本章所讲述的内容,来实现一个网页中常见的信息搜索功能。该实例模拟了音乐网站中常见的歌曲分类搜索,实现的效果如图 2-10 所示。在页面的选择列表框中列出所有的歌曲分类,当用户选择某一类别时,下方显示出所有该类别的歌曲信息。

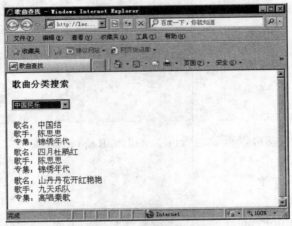

图 2-10 歌曲列表信息

这是一个与数据库交互的应用,我们首先创建应用所需的数据库。在该案例中涉及两张表:歌曲分类表 category 和歌曲表 song,创建脚本如下所示:

```
create table category(
id int identity primary key,
name varchar(20) not null
);
create table song(
id int identity primary key,
name varchar(40) not null,
singer varchar(40),
album varchar(40),
catid int,
foreign key(catid) references category(id)
);
```

在本章项目 Chapter02 中创建两个 JavaBean，分别用于封装歌曲分类信息和歌曲详细信息。这两个类非常简单，Category.java 代码如示例代码 2-9 所示。

示例代码 2-9：实体类 Category.java 源代码

```java
package com.xtgj.j2ee.chapter02.entity;
public class Category {
    private int id; // 音乐类别编号
    private String name;// 音乐类别名称
    // 省略 getter/setter
}
```

Song.java 代码如示例代码 2-10 所示。

示例代码 2-10：实体类 Song.java 源代码

```java
package com.xtgj.j2ee.chapter02.entity;
public class Song {
    private int id; // 歌曲编号
    private String name; // 歌名
    private String singer; // 歌手名
    private String album; // 专辑名
    private int catId; // 所属分类编号
    // 省略 getter/setter
}
```

再创建一个 JavaBean，以封装对歌曲分类信息和歌曲详细信息进行操作的业务逻辑，将该类命名为 Business。在该案例中主要用到两个功能：在页面初始化的时候，在选择列表框中加载所有歌曲分类信息，以及当用户选择歌曲分类时查询该类别的所有歌曲。针对这两个功能，我们在 Business 类中编写了两个业务方法。源代码如示例代码 2-11 所示。

示例代码 2-11：业务逻辑类 Business.java 源代码

```java
package com.xtgj.j2ee.chapter02.user.biz;
import java.sql.Connection;
import java.sql.ResultSet;
import java.sql.Statement;
import java.util.ArrayList;
import java.util.List;
import com.xtgj.j2ee.chapter02.db.DBUtil;
import com.xtgj.j2ee.chapter02.entity.Category;
import com.xtgj.j2ee.chapter02.entity.Song;
```

```java
// 省略 import
public class Business {
    Connection conn = null;
    // 返回所有分类信息
    public List findAllCategories() throws Exception {
        List cats = new ArrayList();
        this.conn = DBUtil.getConn();
        Statement st = conn.createStatement();
        ResultSet rs = st.executeQuery("select * from category");
        while (rs.next()) {
            Category cat = new Category();
            cat.setId(rs.getInt("id"));
            cat.setName(rs.getString("name"));
            cats.add(cat);
        }
        DBUtil.close(rs, st, conn);
        return cats;
    }
    // 根据分类编号查询歌曲
    public List findSongsByCatId(int catId) throws Exception {
        List songs = new ArrayList();
        this.conn = DBUtil.getConn();
        Statement st = conn.createStatement();
        ResultSet rs = st.executeQuery("select * from song where catid="
                + catId);
        while (rs.next()) {
            Song song = new Song();
            song.setId(rs.getInt("id"));
            song.setName(rs.getString("name"));
            song.setSinger(rs.getString("singer"));
            song.setAlbum(rs.getString("album"));
            song.setCatId(rs.getInt("catId"));
            songs.add(song);
        }
        DBUtil.close(rs, st, conn);
        return songs;
    }
}
```

第 2 章 使用 Ajax 改进用户体验

在 Chapter02 项目中创建一个 JSP 文件,命名为 search.jsp,代码如示例代码 2-12 所示。

示例代码 2-12:搜索页面 search.jsp 源代码

```jsp
<%@ page language="java" import="java.util.*" pageEncoding="UTF-8"%>
<%@ page language="java" import="com.xtgj.j2ee.chapter02.entity.*"%>
<%@ page language="java" import="com.xtgj.j2ee.chapter02.user.biz.*"%>
<%
    Business bo = new Business();
    List cats = bo.findAllCategories();
%>
<html>
    <head>
        <title> 歌曲查找 </title>
        <script type="text/javascript">
    var xmlHttp;
    // 创建 XMLHttpRequest 对象的函数
    function createHttpRequest(){
    if(window.XMLHttpRequest){
                    xmlHttp = new XMLHttpRequest();
            }
            else if(window.ActiveXObject){
            try{
                        xmlHttp = new ActiveXObject("Msxml2.XMLHTTP");
                }catch(e){
                try{
                    xmlHttp = new ActiveXObject("Microsoft.XMLHTTP");
                    }catch(e){}
            }
        }
    }
    // 发送请求函数
    function sendRequest(catId){
            if(catId==""){
                    clear();
                    return;
            }
            createHttpRequest();
```

```
            xmlHttp.onreadystatechange = handleResponse;
            var url="Search?catId="+catId;
            xmlHttp.open("get",url,true);
    xmlHttp.send(null);
}
// 处理响应函数
function handleResponse(){
        if(xmlHttp.readyState==4){
    if(xmlHttp.status==200){
      parseXML();
    }
      else{
       alert(" 请求页面有异常 ");
      }
    }
}
// 解析 XML 函数
function parseXML(){
clear();
        var xml = xmlHttp.responseXML;
    var songs = xml.getElementsByTagName("song");
    var tabResult = document.getElementById("results");

    for(var i=0;i<songs.length;i++){
        var name = songs[i].getElementsByTagName("name")[0]
                    .firstChild.data;
    var singer = songs[i].getElementsByTagName("singer")[0]
                    .firstChild.data;
    var album = songs[i].getElementsByTagName("album")[0]
                    .firstChild.data;
            var data =" 歌名:"+name+"<br> 歌手:"+singer
          +"<br> 专集:"+album+"<br>";
        var row = document.createElement("tr");
        var col = document.createElement("td");
        col.innerHTML = data;
        row.appendChild(col);
      tabResult.appendChild(row);

    }
```

```
        }
        // 清空歌曲信息列表
        function clear(){
          var tabResult = document.getElementById("results");
          while(tabResult.childNodes.length>0){
                tabResult.removeChild(tabResult.firstChild);
          }
        }
    </script>
  </head>

  <body>
      <h3>
            歌曲分类搜索
      </h3>
      <select onchange="sendRequest(this.value);">
          <option>
                请选择歌曲类别
          </option>
          <%
                for (int i = 0; i < cats.size(); i++) {
                    Category cat = (Category) cats.get(i);
          %>
          <option value="<%=cat.getId()%>"><%=cat.getName()%>
          </option>
          <%
                }
          %>
      </select>
      <br>
      <br>
      <table cellpaddiing="3">
          <tbody id="results">
          </tbody>
      </table>
  </body>
</html>
```

在页面初始化时,调用 Business 类的 findAllCategories( ) 方法获取数据库中的所有歌曲分类，并加载到一个选择列表框中。当用户选择一个分类时，将调用 sendRequest( ) 方法发送请求，其中处理请求的 Servlet 名为"Search"，并指定 handleResponse( ) 方法处理响应。从处理响应的代码中可以看出，服务器端的处理程序将会把查询到的歌曲信息用 XML 的格式发送到客户端。客户端则综合运用 XML DOM 和 HTML DOM 将响应数据显示在 JSP 页面中。Servlet 的代码如示例代码 2-13 所示。

示例代码 2-13：服务器端处理程序 Search.java 源代码

```java
package com.xtgj.j2ee.chapter02.servlet;

import java.io.IOException;
import java.io.PrintWriter;
import java.util.List;

import javax.servlet.ServletException;
import javax.servlet.http.HttpServlet;
import javax.servlet.http.HttpServletRequest;
import javax.servlet.http.HttpServletResponse;

import com.xtgj.j2ee.chapter02.entity.Song;
import com.xtgj.j2ee.chapter02.user.biz.Business;

public class Search extends HttpServlet{

    public void doGet(HttpServletRequest request,HttpServletResponse response)
                throws ServletException,IOException{

        int catId = Integer.parseInt(request.getParameter("catId"));
        Business bo = new Business();

        List songs = null;
        try{
            // 查询数据库获取歌曲详细信息的集合
            songs = bo.findSongsByCatId(catId);
        }catch(Exception e){
            e.printStackTrace();
            throw new ServletException(" 数据库连接有错 ");

        }
```

```java
                    // 将歌曲信息组织成 XML 格式
StringBuffer xml = new StringBuffer("<songs>");
for(int i=0;i<songs.size();i++){
        Song song = (Song)songs.get(i);
        xml.append("<song>")
                .append("<id>"+song.getId()+"</id>")
                .append("<name>"+song.getName()+"</name>")
                .append("<singer>"+song.getSinger()+"</singer>")
                .append("<album>"+song.getAlbum()+"</album>")
                .append("<catId>"+song.getCatId()+"</catId>")
                .append("</song>");
}
xml.append("</songs>");

response.setContentType("text/xml;charset=UTF-8");
response.setHeader("Cache-Control", "no-cache");
PrintWriter out=response.getWriter();
// 发送 XML 数据
out.println(xml.toString());
out.close();
    }
}
```

在上面的 Servlet 程序中，首先根据客户端发送的请求参数（歌曲分类编号），调用 Business 类的 findSongsByCatId( ) 方法查询相应的歌曲信息，然后将这些数据组织成 XML 格式的数据并发送给客户端。

将该项目部署到 Tomcat 服务器，并启动 Tomcat，然后在浏览器地址栏中输入如下的地址查看运行效果：http://localhost:8080/Chapter02/search.jsp。

2.5 小结

✓ 本章通过从现实中网站的例子引出 Ajax 技术，Ajax 技术通过 JavaScript 发送请求到服务端，并返回结果。
✓ 通过用户注册实例来掌握 Ajax 实现页面刷新的功能。
✓ XMLHttpRequest 对象的方法和属性。

2.6 英语角

category	种类，分类，范畴
onblur	失去焦点事件
asynchronous	异步

2.7 作业

1. 请设计一个使用 Ajax 的案例，并实现该案例。
2. 设计一个鼠标提示信息的案例，效果图如图 2-11 所示。

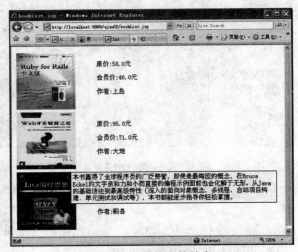

图 2-11　鼠标悬停信息

2.8 思考题

Java EE 的优势主要体现在哪些方面？

2.9　学员回顾内容

1. XMLHttpRequest 对象的作用。
2. XMLHttpRequest 对象的属性和方法。

第 3 章　Ajax 实战技巧

学习目标

- ◆ 了解 Ajax 的应用场合。
- ◆ 理解 Ajax 的实现过程。
- ◆ 掌握 Ajax 的实现技巧。

课前准备

- ◆ 在互联网上查阅资料，知道什么情况下适合使用 Ajax 技术。
- ◆ 进一步掌握 DOM 对象处理的关键思想。

本章简介

在前一章中，我们介绍了 Ajax 的核心技术，相信大家已经掌握了 Ajax 技术的相关方法，例如 XMLHttpRequest 核心对象，传送 XML 格式的数据，使用 DOM 解析 XML 数据及操纵 HTML 文档等。那么，在哪些情况下需要应用 Ajax 技术？本章将通过一些实际案例来帮助大家积累更多的 Ajax 实战经验，体会更多 Ajax 技术的魅力。对 Ajax 应用积累的经验越多，你就越会找到自己的方法来改善 Ajax 应用。

3.1　完成表单验证

对于数据有效性，有一句金玉良言，即防患于未然，杜绝错误的发生。但是如果真的出现了错误，就应该第一时间通知用户。在传统的 Web 应用中必须提交整个页面才能验证数据，或者要依赖复杂的 JavaScript 检验表单。当然有些验证确实很简单，可以使用 JavaScript 编写，但是另外一些验证则不然，完全依靠 JavaScript 编写是办不到的。利用 Ajax，你可以方便地调用服务器端的验证程序，而无需提交整个页面。在多数情况下，这个验证逻辑写起来更简单，测试也更容易。

如果你还不知道应该从哪里开始使用 Ajax，那就不用多想了，就从验证开始吧。本节介绍的例子是最常见的验证之一——日期验证。这个例子的 HTML 页面很简单，运行效果如图 3-1 和图 3-2 所示。

第 3 章 Ajax 实战技巧

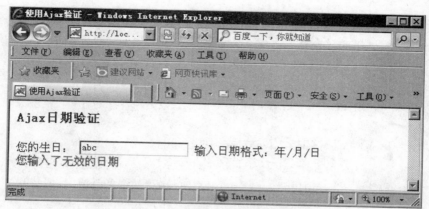

图 3-1 日期验证失败的页面

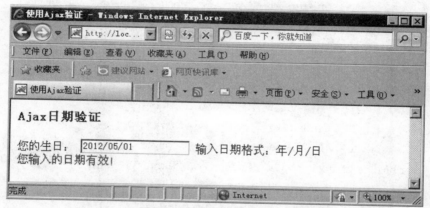

图 3-2 日期验证成功的页面

其中有一个文本输入框,用于输入一个日期值,该文本框的 onblur() 事件会触发验证函数 validate(),在验证函数中首先调用标准的 createXMLHttpRequest() 函数创建 XMLHttpRequest 对象,然后把用户输入的日期发送到 ValidateServlet 进行验证,handleResponse() 函数从服务器获取验证结果,然后委托给 setMessage() 函数,setMessage() 函数会检查验证结果值以确定用什么颜色来显示消息。具体实现步骤如下:

首先,在 Eclipse 中新建一个 Web 项目,命名为 Chapter03。

然后,在该项目中新建一个 html 文件,命名为 validate.html,在该文件中编写代码如示例代码 3-1 所示。

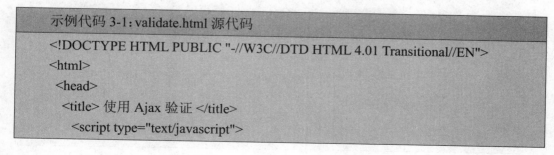

示例代码 3-1:validate.html 源代码
```
<!DOCTYPE HTML PUBLIC "-//W3C//DTD HTML 4.01 Transitional//EN">
<html>
  <head>
    <title> 使用 Ajax 验证 </title>
      <script type="text/javascript">
```

```
var xmlHttp;
// 创建 XMLHttpRequest 请求对象
function createXMLHttpRequest(){
    if(window.XMLHttpRequest){
        xmlHttp = new XMLHttpRequest();
    }
    else if(window.ActiveXObject){
        try{
            xmlHttp = new ActiveXObject("Msxml2.XMLHTTP");
        }catch(e){
            try{
                xmlHttp = new ActiveXObject("Microsoft.XMLHTTP");
            }catch(e){}
        }
    }
}
// 验证函数（发送请求）
function validate(){
    createXMLHttpRequest();
    var birthday = document.getElementById("birthday").value;
    var url = "Validate?birthday="+birthday;
    xmlHttp.open("GET",url,true);
    xmlHttp.onreadystatechange = handleResponse;
    xmlHttp.send(null);
}
// 处理响应信息
function handleResponse(){
    if(xmlHttp.readyState==4){
        if(xmlHttp.status==200){
            var xml = xmlHttp.responseXML;
            var msg = xml.getElementsByTagName("message")[0]
                .firstChild.nodeValue;
            var valid = xml.getElementsByTagName("valid")[0]
                .firstChild.codeValue;
            setMessage(msg,valid);
        }
    }
}
```

```
            // 显示验证消息
            function setMessage(message,isValid){
                    var msgDiv = document.getElementById("dateMessage");
                    var fontColor = "red";
                    if(isValid=="true"){
                fontColor = "green";
                    }
                    msgDiv.innerHTML = "<font color="+fontColor+">"
                      +message+"</font>";
            }
    </script>
  </head>

  <body>
  <h3>Ajax 日期验证 </h3>
  您的生日：
  <input type="text" id="birthday" onblur="validate();">
  输入日期格式：年 / 月 / 日
  <div id ="dateMessage">
  </div>
  </body>
</html>
```

接着，编写 Servlet 类 ValidateServlet，用于处理前端页面发来的请求。其源代码如示例代码 3-2 所示。

示例代码 3-2：ValidateServlet.java 源代码

```java
package com.xtgj.j2ee.chapter03.xtgj.servlets;

import java.io.IOException;
import java.io.PrintWriter;
import java.text.DateFormat;
import java.text.ParseException;
import java.text.SimpleDateFormat;

import javax.servlet.ServletException;
import javax.servlet.http.HttpServlet;
import javax.servlet.http.HttpServletRequest;
```

```java
import javax.servlet.http.HttpServletResponse;
public class ValidateServlet extends HttpServlet {
    public void doGet(HttpServletRequest request, HttpServletResponse response)
                throws ServletException, IOException {
        request.setCharacterEncoding("utf-8");
        // 获取用户输入的日期
        String date = request.getParameter("birthday");
        // 验证日期有效性
        boolean valid = validate(date);
        String msg = " 您输入了无效的日期 ";
        if (valid == true) {
            msg = " 您输入的日期有效！ ";
        }
        // 设置响应格式和字符集
        response.setContentType("text/xml;charset=utf-8");
        response.setHeader("Cache-Control", "no-cache");
        PrintWriter out = response.getWriter();
        // 输出响应数据
        out.println("<response>");
        out.println("<message>" + msg + "</message>");
        out.println("<valid>" + valid + "</valid>");
        out.println("</response>");
        out.close();
    }

    // 验证日期方法
    private boolean validate(String date) {
        boolean valid = true;
        DateFormat formatter = new SimpleDateFormat("yyyy/MM/dd");
        try {
            formatter.parse(date);
        } catch (ParseException e) {
            valid = false;
        }
        return valid;
    }
}
```

最后，在 web.xml 中配置该 Servlet。其源代码片段如示例代码 3-3 所示。

示例代码 3-3：web.xml 配置源代码片段

```xml
<servlet>
    <servlet-name>ValidateServlet</servlet-name>
    <servlet-class>com.xtgj.j2ee.chapter03.xtgj.servlets.ValidateServlet</servlet-class>
</servlet>
<servlet-mapping>
    <servlet-name>ValidateServlet</servlet-name>
    <url-pattern>/Validate</url-pattern>
</servlet-mapping>
```

将完成的项目部署到 Tomcat，启动 Tomcat 后在浏览器地址栏中输入如下的地址查看运行效果：http://localhost:8080/Chapter03/validate.html。

3.2 动态加载列表框

Web 应用通常使用"向导工具"设计原则来构建，即每个屏幕要求用户输入少量的信息，每个后续页的数据都依据前一页的输入来创建。对于某些情况，这个设计模式非常有用，可以使用户以一种有序的方式完成任务。遗憾的是，在 Ajax 技术出现之前，当基于用户输入修改页面的某些部分时，动态地更新页面而不刷新整个页面是很难办到的，甚至根本不可能。

避免页面刷新的一种技术是在页面上隐藏数据，并在需要时再显示它们。例如：选择框 B 的值要根据选择框 A 中选值来填写，此时 B 的所有可取值就可以放在隐藏的选择框中，当选择框 A 中的所选值变化时，JavaScript 可以确定要显示哪一个隐藏的选择框，并将它置为可见，同时把前一个选择框置为不可见。这是一种很有用的技术，但它们只是在有限的情况下可用，即页面中仅限于根据用户输入对有限的选择进行修改，而且这样的选择必须相对较少。

假设你在构建一个在线的汽车分类广告服务。某人想购买汽车，指定了车型年份、品牌和车型，来搜索他想买的汽车。为了避免用户输入错误，并减少验证次数，决定车型年份、品牌和车型等输入字段都使用选择框，如果车型年份和品牌选择框中的选择发生变化时，相应的车型选择框中的值也应该同时更新。

每年每个品牌都会生产多种车型，那么年份、品牌和车型的组合数是惊人的，只用 JavaScript 来填写隐藏选择框是不可能的。使用 Ajax 技术可以轻松地解决这个问题。年份或品牌选择框中的选择每次有变化时，会向服务器发出异步请求，要求得到该车型年份特定品牌的车型列表。服务器根据请求的年份和品牌来找到相应的车型列表，并把它们打包在一个 XML 文件中，返回给浏览器。浏览器负责解析 XML，并将获得的车型列表填写到车型选择框中。

下面我们来具体实现这个案例。本例实现效果如图 3-3 所示。

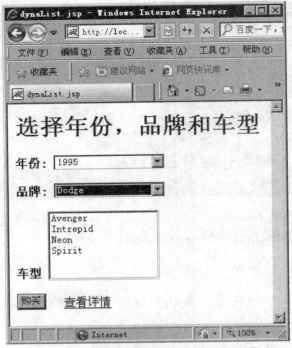

图 3-3 根据年份和品牌级联选择汽车型号

在这个页面中显示了一个表单,其中有三个列表框,分别是年份、品牌和车型。当用户选择了年份或者品牌,在下面的车型列表框中将会动态显示相应的车型列表,以供用户做进一步选择。由于本案例只是演示动态加载列表框,所以界面中的"购买"按钮和"查看详情"超链接并没有实现什么功能。这是一个与数据库交互的应用,其中涉及一张表,即汽车表 car。我们在 SQL Server 数据库管理系统中创建数据库 test,并在数据库中建立这张表,同时插入一些测试数据,如示例代码 3-4 所示。

示例代码 3-4:数据库源代码片段

```
create table car(
id int identity primary key,
year int not null,
make varchar(20) not null,
model varchar(20) not null
);

insert into car(year,make,model) values(2006,'Dodge','Charger');
insert into car(year,make,model) values(2006,'Dodge','Magnum');
insert into car(year,make,model) values(2006,'Dodge','Ram');
insert into car(year,make,model) values(2006,'Dodge','Viper');
insert into car(year,make,model) values(1995,'Dodge','Avenger');
```

```
insert into car(year,make,model) values(1995,'Dodge','Intrepid');
insert into car(year,make,model) values(1995,'Dodge','Neon');
insert into car(year,make,model) values(1995,'Dodge','Spirit');
```

现在，我们一起看一看上述功能的实现步骤。首先，在 Chapter03 项目中创建一个 JavaBean，其中封装了操作数据库的方法，以提供界面所需的数据，如示例代码 3-5 所示。

示例代码 3-5：业务逻辑 DBOperator.java 源代码片段

```java
package com.xtgj.j2ee.chapter03.car.beans;

import java.sql.Connection;
import java.sql.PreparedStatement;
import java.sql.ResultSet;
import java.sql.Statement;
import java.util.ArrayList;
import java.util.List;

import com.xtgj.j2ee.chapter02.db.DBUtil;

public class DBOperator {
    // 查找所有品牌列表
    public List findMakes() throws Exception {
        List list = new ArrayList();
        Connection cn = getConnection();
        Statement st = cn.createStatement();
        ResultSet rs = st.executeQuery("select distinct make from car");
        while (rs.next()) {
            list.add(rs.getString(1));
        }
        rs.close();
        st.close();
        cn.close();
        return list;
    }

    // 根据年份和品牌查找相应的车型列表
    public List findModels(int year, String make) throws Exception {
```

```java
        List list = new ArrayList();
        Connection cn = getConnection();
        String sql = "select model from car where year=? and make=?";
        PreparedStatement pst = cn.prepareStatement(sql);
        pst.setInt(1, year);
        pst.setString(2, make);
        ResultSet rs = pst.executeQuery();
        while (rs.next()) {
            list.add(rs.getString(1));
        }
        rs.close();
        pst.close();
        cn.close();
        return list;
}

// 查找所有年份列表
public List findYears() throws Exception {
        List list = new ArrayList();
        Connection cn = getConnection();
        Statement st = cn.createStatement();
        ResultSet rs = st.executeQuery("select distinct year from car");
        while (rs.next()) {
            list.add(new Integer(rs.getInt(1)));
        }
        rs.close();
        st.close();
        cn.close();
        return list;
}

public List findModels(String modelPrefix) throws Exception {
        List list = new ArrayList();
        Connection cn = getConnection();
        String sql = "select model from car where model like '" + modelPrefix
                + "%' group by model";
        Statement st = cn.createStatement();
        ResultSet rs = st.executeQuery(sql);
```

```
            while (rs.next()) {
                    list.add(rs.getString(1));
            }
            rs.close();
            st.close();
            cn.close();
            return list;
    }

    // 获取数据库连接
    private Connection getConnection() throws Exception {
            return DBUtil.getConn();
    }
```

接着,创建一个 JSP 文件,把它命名为 dynaList.jsp,如示例代码 3-6 所示。

示例代码 3-6:视图 dynaList.jsp 源代码片段

```
package com.xtgj.j2ee.chapter03.car.beans;

import java.sql.Connection;
import java.sql.PreparedStatement;
import java.sql.ResultSet;
import java.sql.Statement;
import java.util.ArrayList;
import java.util.List;

import com.xtgj.j2ee.chapter02.db.DBUtil;

public class DBOperator {
    // 查找所有品牌列表
    public List findMakes() throws Exception {
            List list = new ArrayList();
            Connection cn = getConnection();
            Statement st = cn.createStatement();
            ResultSet rs = st.executeQuery("select distinct make from car");
            while (rs.next()) {
                    list.add(rs.getString(1));
            }
```

```java
            rs.close();
            st.close();
            cn.close();
            return list;
    }

    // 根据年份和品牌查找相应的车型列表
    public List findModels(int year, String make) throws Exception {
            List list = new ArrayList();
            Connection cn = getConnection();
            String sql = "select model from car where year=? and make=?";
            PreparedStatement pst = cn.prepareStatement(sql);
            pst.setInt(1, year);
            pst.setString(2, make);
            ResultSet rs = pst.executeQuery();
            while (rs.next()) {
                    list.add(rs.getString(1));
            }
            rs.close();
            pst.close();
            cn.close();
            return list;
    }

    // 查找所有年份列表
    public List findYears() throws Exception {
            List list = new ArrayList();
            Connection cn = getConnection();
            Statement st = cn.createStatement();
            ResultSet rs = st.executeQuery("select distinct year from car");
            while (rs.next()) {
                    list.add(new Integer(rs.getInt(1)));
            }
            rs.close();
            st.close();
            cn.close();
            return list;
    }
```

第 3 章　Ajax 实战技巧　　61

```java
    public List findModels(String modelPrefix) throws Exception {
        List list = new ArrayList();
        Connection cn = getConnection();
        String sql = "select model from car where model like '" + modelPrefix
                + "%' group by model";
        Statement st = cn.createStatement();
        ResultSet rs = st.executeQuery(sql);
        while (rs.next()) {
            list.add(rs.getString(1));
        }
        rs.close();
        st.close();
        cn.close();
        return list;
    }

    // 获取数据库连接
    private Connection getConnection() throws Exception {
        return DBUtil.getConn();
    }
}
```

然后，创建服务器端处理程序，类名为 SearchModel，如示例代码 3-7 所示。

示例代码 3-7：服务器端处理程序 SearchModel.java 源代码片段

```java
package com.xtgj.j2ee.chapter03.servlets;

import java.io.IOException;
import java.io.PrintWriter;
import java.util.Iterator;
import java.util.List;

import javax.servlet.ServletException;
import javax.servlet.http.HttpServlet;
import javax.servlet.http.HttpServletRequest;
import javax.servlet.http.HttpServletResponse;
```

```java
import com.xtgj.j2ee.chapter03.car.beans.DBOperator;

public class SearchModel extends HttpServlet {
    public void doGet(HttpServletRequest request,HttpServletResponse response)
            throws ServletException,IOException{
        // 获取年份和品牌
        int year = Integer.parseInt(request.getParameter("year"));
        String make = request.getParameter("make");
        // 根据年份和品牌查询车型
        DBOperator db = new DBOperator();
        List list = null;
        try{
            list = db.findModels(year, make);
        }
        catch(Exception e){
            throw new ServletException(" 数据库操作错误 !");
        }
        // 将车型打包为 XML 格式的数据
        StringBuffer xml = new StringBuffer("<models>");
        for(Iterator it = list.iterator();it.hasNext();){
            String model = (String)it.next();
            xml.append("<model>").append(model).append("</model>");
        }
        xml.append("</models>");
        // 设置响应格式的字符集
        response.setContentType("text/xml;charset=utf-8");
        response.setHeader("Cache-Control", "no-cache");
        PrintWriter out = response.getWriter();
        // 发送响应数据
        out.print(xml.toString());
        out.close();
    }
    public void doPost(HttpServletRequest request
       ,HttpServletResponse response)
            throws ServletException,IOException{
        doGet(request,response);
    }
}
```

最后，在 web.xml 中对所创建的 Servlet 进行如示例代码 3-8 所示的配置。

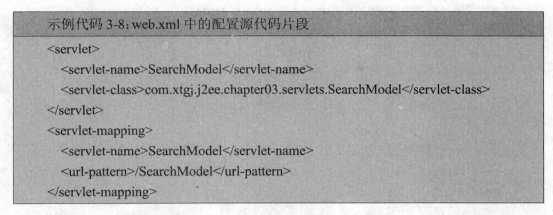

示例代码 3-8：web.xml 中的配置源代码片段
```
<servlet>
    <servlet-name>SearchModel</servlet-name>
    <servlet-class>com.xtgj.j2ee.chapter03.servlets.SearchModel</servlet-class>
</servlet>
<servlet-mapping>
    <servlet-name>SearchModel</servlet-name>
    <url-pattern>/SearchModel</url-pattern>
</servlet-mapping>
```

将完成的项目部署到 Tomcat，启动 Tomcat 后在浏览器地址栏中输入如下的地址查看运行效果：http://localhost:8080/Chapter03/dynaList.jsp。

3.3 实现自动完成

我们遇到的最受欢迎的功能之一就是自动完成。著名的搜索网站 Google 提供的自动完成功能确实让人赞叹不已，当用户在搜索输入框中输入文字时，会及时显示一个下拉区，并提供一些最有可能的答案让用户进行快速选择，在下拉区中甚至还能使用上、下箭头键。为给定项提供了一些结果，用户就能更清楚地知道具体完成搜索时可能会得到的结果。

本章最后一个案例，我们来实现自动完成功能。我们的例子所实现的功能没有 Google 的那么丰富，但是我们可以从中了解到利用 Ajax 可以做些什么。

在本例中，当用户在文本框中输入待查询的车型名称，页面中将根据用户输入的文字信息及时给出提示列表，以帮助用户进行快速选择，这些提示信息来源于数据库。为了方便起见，本例仍然使用上一个案例所创建的项目 Chapter03，以及该案例中所创建的数据表 car。

本案例的实现效果如图 3-4 所示。

图 3-4　Ajax 自动完成

首先,我们在 Chapter03 项目中找到 DBOperator 类,在其中添加一个方法,该方法根据车型名称的前缀查询 car 表中匹配的车型列表,其源代码如示例代码 3-9 所示。

示例代码 3-9:DBOperator.java

```java
package com.xtgj.j2ee.chapter03.car.beans;

import java.sql.Connection;
import java.sql.PreparedStatement;
import java.sql.ResultSet;
import java.sql.Statement;
import java.util.ArrayList;
import java.util.List;

import com.xtgj.j2ee.chapter02.db.DBUtil;

public class DBOperator {
    // 查找所有品牌列表
    public List findMakes(String modelPrefix) throws Exception {
        List list = new ArrayList();
        Connection cn = getConnection();
        Statement st = cn.createStatement();
        String sql = "select model from car where model like '" + modelPrefix
                + "%' group by model";
        ResultSet rs = st.executeQuery(sql);

        while (rs.next()) {
            list.add(rs.getString(1));
        }
        rs.close();
        st.close();
        cn.close();
        return list;
    }

    // 获取数据库连接
    private Connection getConnection() throws Exception {
        return DBUtil.getConn();
    }
```

}

接下来创建搜索页面,命名为 autoComplete.jsp,如示例代码 3-10 所示。

示例代码 3-10:视图 autoComplete.jsp 源代码片段

```jsp
<%@ page language="java" import="java.util.*" pageEncoding="GBK"%>
<!DOCTYPE HTML PUBLIC "-//W3C//DTD HTML 4.01 Transitional//EN">
<html>
    <head>
            <title> 自动完成 </title>
            <style type="text/css">
.mouseOver {
    background: #708090;
    color: #FFFAFA;
}

.mouseOut {
    background: #FFFAFA;
    color: #000000;
}
</style>
            <script type="text/javascript">
    var xmlHttp;
    var inputField;
    var popupDiv;
    var popupBody;
    // 创建 XMLHttpRequest 对象
    function createXMLHttpRequest(){
            if(window.XMLHttpRequest){
                    xmlHttp = new XMLHttpRequest();
            }
            else if(window.ActiveXObject){
                    try{
                            xmlHttp = new ActiveXObject("Msxml2.XMLHTTP");
                    }catch(e){
                            try{
                                    xmlHttp = new ActiveXObject("Microsoft.XMLHTTP");
                            }catch(e){}
```

```
            }
        }
}
// 发送匹配请求函数
function findModels(){
        inputField = document.getElementById("model");
        popupDiv = document.getElementById("popup");
        popupBody = document.getElementById("popupBody");

        if(inputField.value==""){
                clearModels();
                return;
        }

        createXMLHttpRequest();
        var url = "AutoComplete";
        var queryStr = "model="+inputField.value;
        xmlHttp.open("post",url,true);
        xmlHttp.onreadystatechange = handleResponse;
        xmlHttp.setRequestHeader("Content-Type"
            ,"application/x-www-form-urlencoded");
        xmlHttp.send(queryStr);
}
// 清除自动完成行
function clearModels(){
        while(popupBody.childNodes.length>0){
                popupBody.removeChild(popupBody.firstChild);
        }
        popupDiv.style.border = "none";
}
// 处理响应数据
function handleResponse(){
        if(xmlHttp.readyState == 4){
                if(xmlHttp.status == 200){
                        setModels();
                }
        }
}
```

```javascript
// 生成自动完成显示行
function setModels(){
        clearModels();
        var models = xmlHttp.responseXML
                    .getElementsByTagName("model");
        if(models.length == 0)return;

        setOffsets();
        var row,cell,text;
        for(var i=0;i<models.length;i++){
                text = document.createTextNode(models[i]
                        .firstChild.nodeValue);
                cell = document.createElement("td");
                row = document.createElement("tr");

cell.onmouseover = function(){this.className="mouseOver;"};
cell.onmouseout = function(){this.calssName="mouseOut;"};
cell.onclick = function(){populateModel(this);};

                cell.appendChild(text);
                row.appendChild(cell);
                popupBody.appendChild(row);
        }
}
// 写入输入框
function populateModel(cell){
inputField.value = cell.firstChild.nodeValue;
clearModels();
 }
// 设置自动提示框的显示位置
function setOffsets(){
        var width = inputField.offsetWidth;
        var left = getLeft(inputField);
        var top = getTop(inputField)+inputField.offsetHeight;

        popupDiv.style.border = "black 1px solid";
        popupDiv.style.left = left+"px";
        popupDiv.style.top = top+"px";
```

```
            popupDiv.style.width = width+"px";
    }
    // 获取指定元素在页面中的宽度起始位置
    function getLeft(elem){
        var offset = elem.offsetLeft;
        if(elem.offsetParent!=null)
            offset += getLeft(elem.offsetParent);
        return offset;
    }
    // 获取指定元素在页面中的高度起始位置
    function getTop(elem){
        var offset = elem.offsetTop;
        if(elem.offsetParent!=null)
            offset += getTop(elem.offsetParent);
        return offset;
    }
    </script>
    </head>

    <body>
        <h2>
            Ajax 自动完成功能
        </h2>
        车型：
        <input id="model" size="20" style="height: 20;"
            onkeyup="findModels();">
        <input type="button" value=" 搜索 ">

        <div id="popup" style="position: absolute;">
            <table width="100%" bgcolor="#fffafa" cellspacing="0" cellpadding="0">
                <tbody id="popupBody">
                </tbody>
            </table>
        </div>
    </body>
</html>
```

在该页面中，一旦用户在文本框中输入文字，即引发"onkeyup"事件，并调用 findModels()

函数。在该函数中先初始化一些全局变量，然后借助 Ajax 提交请求，同时提交用户已输入的信息。在处理响应的函数 handleResponse() 中调用 setModels() 函数以完成信息自动提示的功能。

下面我们来实现服务器端 Servlet 程序。本例使用了 POST 提交方式，在 autoComplete.java 的 doPost() 方法中，首先获取请求数据，即需要匹配车型的字符串前缀，然后调用数据库操作的 JavaBean 完成数据库的查询，最后将获得的数据以 XML 格式发送回客户端。该 Servlet 的代码如示例代码 3-11 所示。

示例代码 3-11：服务器端处理程序 AutoComplete.java 源代码片段

```java
package com.xtgj.j2ee.chapter03.servlets;

import java.io.IOException;
import java.io.PrintWriter;
import java.util.List;

import javax.servlet.ServletException;
import javax.servlet.http.HttpServlet;
import javax.servlet.http.HttpServletRequest;
import javax.servlet.http.HttpServletResponse;

import com.xtgj.j2ee.chapter03.car.beans.DBOperator;

public class AutoComplete extends HttpServlet {

    public void doPost(HttpServletRequest request, HttpServletResponse response)
            throws ServletException, IOException {
        request.setCharacterEncoding("utf-8");
        // 获取需要匹配的字符串前缀
        String prefix = request.getParameter("model");
        // 查询匹配的车型列表
        DBOperator db = new DBOperator();
        List models = null;
        try {
            models = db.findMakes(prefix);
        } catch (Exception e) {
            throw new ServletException(" 数据库错误 ");
        }
        // 设置响应格式和字符集
```

```java
response.setContentType("text/xml;charset=utf-8");
response.setHeader("Cache-Control", "no-cache");
PrintWriter out = response.getWriter();
// 以 XML 格式输出相应数据
out.println("<response>");
for (int i = 0; i < models.size(); i++) {
        String model = (String) models.get(i);
        out.println("<model>" + model + "</model>");
}
out.println("</response>");
out.close();
    }
}
```

在 web.xml 对以上的 Servlet 进行如下配置:

示例代码 3-12: web.xml 配置源代码片段

```xml
<servlet>
   <servlet-name>AutoComplete</servlet-name>
   <servlet-class>com.xtgj.j2ee.chapter03.servlets.AutoComplete</servlet-class>
</servlet>
<servlet-mapping>
   <servlet-name>AutoComplete</servlet-name>
   <url-pattern>/AutoComplete</url-pattern>
</servlet-mapping>
```

将完成的项目部署到 Tomcat，启动 Tomcat 后在浏览器地址栏中输入如下地址查看运行效果：http://localhost:8080/Chapter03/autoComplete.jsp。

3.4 小结

✓ 本章提供了几个例子，展示了如何应用 Ajax 技术改善用户体验。相信这些例子可以帮助你在短时间内体验 Ajax 的方便、灵活、人性化的交互方式。

✓ 在许多情况下，可以把 Ajax 技术应用到现有的应用中，以 Ajax 请求取代完全页面刷新，从而与服务器无缝地通信，并更新页面内容。

✓ Ajax 技术也可以在开始设计新的 Web 应用时就采用。

3.5 英语角

validate	验证
dynamic	动态
refresh	刷新
complete	完成

3.6 作业

实现一个网上书店中图书分类级联菜单的设计。菜单项应该来自于数据库,数据库中包含图书主分类表和次分类表,当页面加载时显示图书主分类菜单项列表,当用户点击某一个主菜单项时,在该项菜单的下方展开并显示相应的次分类菜单项列表,当用户再次点击主菜单项时则收缩子菜单项列表。

3.7 思考题

在哪些场合适合使用 Ajax 技术?如何实现?

3.8 学员回顾内容

1. Ajax 案例的设计思路与实现过程。
2. Ajax 实现技巧。

第 4 章 使用 Hibernate 完成对象持久化

学习目标

- ◆ 了解 Hibernate 的运行原理。
- ◆ 了解 Hibernate 的优势。
- ◆ 掌握使用 Hibernate 的基本步骤。
- ◆ 会使用工具简化 Hibernate 操作。

课前准备

通过互联网搜索有关 ORM 的相关资料

本章简介

Web 的出现改变了我们的生活，从此人们对数据和信息越来越依赖，直到今天，Web 开始指导我们的吃穿住用行。财富和价值的创造方式也变得与往日不同。从互联网行业的火爆到企业应用开发领域内 B/S 大行其道，浏览器悄然间成了左右这个世界的重要角色。

到本章为止，我们经历了从静态页面到 JSP，再到 JSP+Servlet+JavaBean 的学习过程，又在前面学习了 Struts 技术。Java EE 技术的学习曲线也在一定程度上反应技术发展的过程。纯 JSP 盛行的时代是乱世，那时候编写的 HTML 和 JSP 混用代码很少有存活到今天的。业务逻辑越来越复杂，JavaBean 和 Servlet 开始广泛应用，这个世界里开始散发着文明的气息，事情开始变得有秩序。Struts 出现的意义大致相当于《拿破仑法典》的颁布，框架的广泛使用摒除了杂乱和无序，还提供给人们各种方便。EJB 王朝没落的时候，三层架构 +Struts MVC 开始了温和的统治，这个新的领袖虽然没有王冠（非 JCP 委员会），也没有权杖（不是 JSR 标准），但成为了 Java Web 应用开发事实标准，得到了天下。

江水滚滚东去，技术前进的脚步也从未停歇。技术的前进刺激了需求，需求激增又对技术提出了更高的要求。三层架构和 Struts 虽然定义了整体的程序框架，但程序员还有很多琐碎的工作要处理，当软件需求量越来越大的时候，我们需要写太多繁琐的重复性代码。这是必须要承担的工作吗？聪明的程序员总希望用更有效、更稳妥的方法解决问题。

客观上我们需要提高程序编码的效率和灵活性，一方面应对越来越复杂的需求，一方面把我们从繁琐的编码工作中解脱出来，使我们能够把精力集中在真正需要思考的事情上。在软件大规模生产中，如果在某一环节提高了生产率，对企业和行业的贡献将是非常巨大的。

在 2001 年以前，Gavin King 是一个名不见经传的小人物，平凡得同你我一样。他与众不同之处在于他的激情洋溢、脾气倔强和永不言败。在工作过程中他觉得总是花很多时间进行重复性的编码效率不高，他相信自己可以找到一个更好的方法提高数据访问层的编码效率。两年后，他做到了。他开发出了全世界最流行的"对象 - 关系映射"工具：Hibernate。那个对 SQL 和数据库一窍不通的小伙子居然成了全世界 Java 数据持久化解决方案的领导者。热情和坚持帮助 Gavin 成了国王。

本章我们将关注 Gavin King 的 Hibernate：当前最优秀的"对象—关系"映射工具，主流的持久化解决方案。

4.1 为什么需要 Hibernate

2016 年，北京某中心 Y2 学员小周在做毕业设计项目的时候，发现反反复复地写 DAO 层代码太麻烦了。如示例代码 4-1 所示，是一个租房网的租房业务逻辑类中封装的录入房屋信息的方法。他负责的模块每个表都少则十几个字段，多则几十个字段，这种重复性的编码工作没有任何创造性，而且容易出错。

示例代码 4-1：早期的 DAO

```java
public void insertFwxx(FWXX fwxx) {
    openConn();
    String sql = "";
    sql="insert into TBL_FWXX(uid,jdid,lxid,shi,ting,fwxx,zj,title,date,telephone,lxr) values(?,?,?,?,?,?,?,?,?,?,?)";
    try {
        pstmt = conn.prepareStatement(sql);
        pstmt.setInt(1, fwxx.getUid());// uid
        pstmt.setInt(2, fwxx.getJdid());// 街道表 id
        pstmt.setInt(3, fwxx.getLxid());// 房屋类型表 id
        pstmt.setInt(4, fwxx.getShi());// 几室
        pstmt.setInt(5, fwxx.getTing());// 几厅
        pstmt.setString(6, fwxx.getFwxx());// 租房信息
        pstmt.setInt(7, fwxx.getZj()); // 月租金
        pstmt.setString(8, fwxx.getTitle());// 标题
        pstmt.setString(9, (new SimpleDateFormat("yyyy-MM-dd HH:mm:ss"))
            .format(new Date()));// 插入时间
        pstmt.setString(10, fwxx.getTelephone());
        pstmt.setString(11, fwxx.getLxr());
        pstmt.executeUpdate();
    } catch (SQLException e) {
```

```
            e.printStackTrace();
        }
        closeAll();
    }
```

这样的编码工作太繁琐了,占用了他大部分的开发时间。不过为了实现一个功能强大、让自己满意的系统,他坚持了下来。在实现了所有功能之后,他觉得肯定有更聪明的方法!在老师的指导下,他自学了课本上没有介绍过的反射技术,深入学习了 JDBC 中 metadata 相关的知识,一个方案在他头脑中成形了。小周是一个事情没做好吃饭睡觉都不踏实的人。短短 3 天时间,他实现了一个通用的 DAO 类。从而再也不用理会那些繁琐的字段名和 pstmt 的 getter/setter 方法了。

小周马上将他的经验推广给小组其他成员。这极大地推进了项目的进度。小组获得了最佳团队的荣誉,小周也被评为最佳个人。

2001 年,Gavin 遇到过类似的情况。他觉得当时数据库访问的代码开发效率太低了,他觉得可以开发出一套更好的数据库访问框架,把项目开发的时间大大缩短。有了初步的想法后,他决定开始行动!他开发 Hibernate 的第一件事情是去街上买了一本 SQL 入门的书。两年后,他做到了。当我们对现实感到不满意的时候,新的技术就会出现。

如今我们开发的大多数软件产品都是基于数据库的,数据访问层的繁琐代码严重影响了我们的开发效率。企业做软件是在和时间、成本赛跑。福特汽车悬赏 2.5 万元,征求有办法让每一台汽车上节省一个螺钉和螺帽的人。软件行业又何尝不是这样呢?

Hibernate 的价值在于,对 JDBC 访问数据库的代码做了封装,大大简化了数据访问层繁琐的重复性代码。到本章结束的时候,我们可以看到:使用 Hibernate 之后的 DAO 层代码和之前的相比清爽了多少!

4.2　Hibernate 是什么

概括地说,Hibernate 是一个优秀的 Java 持久化解决方案,是当今主流的"对象—关系"映射工具。什么是持久化解决方案,什么又是"对象—关系"映射呢?请看下文。

4.2.1　持久化

程序运行的时候,有些程序数据保存在内存中,当程序退出后,这些数据就不复存在了,所以,我们称这些数据的状态为瞬时的(Transient)。有些数据,在程序退出后,还以文件等形式保存在存储设备(硬盘、光盘等)中,我们称这些数据的状态是持久的(Persistent)。

持久化是将程序中数据在瞬时状态和持久状态间转换的机制。持久化概念图如图 4-1 所示。

第 4 章 使用 Hibernate 完成对象持久化

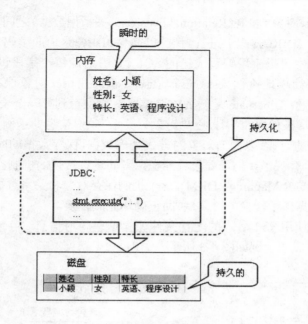

图 4-1 持久化概念

JDBC 就是一种持久化机制。将程序数据直接保存成文本文件也是持久化机制的一种实现。但我们常用的是将程序数据保存到数据库中。

在三层结构中，DAO 层（数据访问层）有时候也称为持久化层，如图 4-2 所示。因为这一层承担的主要工作就是将数据保存到数据库中或把数据从数据库中读取出来。

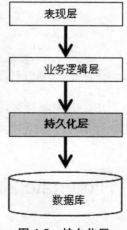

图 4-2 持久化层

4.2.2 "对象—关系"映射

我们多以面向对象的方式组织程序，瞬时的数据也多以对象的形式存在，而持久的数据多保存在关系型数据库中。所以,在通常情况下,持久化要完成的操作就是把对象保存到关系型数据库中,或者把关系型数据库中的数据读取出来,以对象的形式封装。所以,我们的持久化工作主要在O（Object，对象）和R（Relational Database，关系型数据）之间完成。

Hibernate就是在JDBC方式上进行封装，以简化JDBC繁琐的编码工作。使用Hibernate将对象保存到数据库中再也不用编写大量的SQL语句，对应每个字段设置PreparedStatement中参数的值。只需要简单地执行session.save(user)，即可把user对象保存到数据库对应的表中。这就涉及一个问题：Hibernate是怎么知道user对象保存到哪一个表中，user对象中的每个属性又对应到数据库表的哪个字段呢？"对象—关系"映射信息的示例如图4-3所示。

我们在编写程序的时候，以面向对象的方式处理数据，保存数据的时候，却以关系型数据库的方式存储。所以，客观上我们需要一种能在两者间进行转换的机制。这样的机制称为"对象—关系"映射机制(O/R Mapping, ORM)，这个机制保存对象和关系数据库表的映射信息，当数据在对象和关系数据库中转换的时候，协助正确地完成转换。

在Hibernate中，使用xml格式的配置文件保存这些映射信息。在4.4节我们即将学到。"对象—关系"映射信息的示例如图4-3所示。

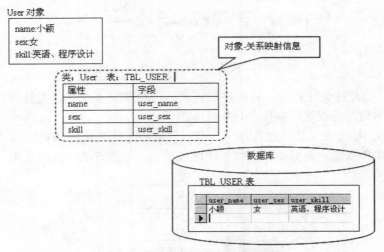

图4-3 "对象—关系"映射信息

综上所述，Hibernate是一个基于JDBC的主流持久化框架，是一个优秀的ORM实现。它能很大程度地简化DAO层的编码工作，下面我们就来看看Hibernate如何发挥它的神奇魔力。

4.3 使用Hibernate的基本步骤

4.3.1 准备工作

本章案例主要实现租房信息管理系统的业务逻辑，用到的数据库可以是SQL Server数据库或其他。我们在SQL Server数据库管理系统中创建名为"zf"的数据库，并创建如下所示的数据表，读者可以根据注释内容理解数据库表的基本结构。

```sql
-- 房屋类型表
CREATE TABLE [dbo].[TBL_FWLX] (
    [lxid] [int] NOT NULL ,                          -- 类型编号
    [fwlx] [varchar] (50) COLLATE Chinese_PRC_CI_AS NULL  -- 类型名称
) ON [PRIMARY]
GO

-- 房屋信息表
CREATE TABLE [dbo].[TBL_FWXX] (
    [fwid] [int] IDENTITY (1, 1) NOT NULL ,          -- 房屋编号
    [uid] [int] NULL ,                               -- 发布用户编号
    [jdid] [int] NULL ,                              -- 街道编号
    [lxid] [int] NULL ,                              -- 类型编号
    [shi] [int] NULL ,                               -- 几室
    [ting] [int] NULL ,                              -- 几厅
    [fwxx] [varchar] (2000) COLLATE Chinese_PRC_CI_AS NULL ,  -- 房屋信息描述
    [title] [varchar] (70) COLLATE Chinese_PRC_CI_AS NOT NULL ,  -- 标题
    [zj] [varchar](70) NULL ,                        -- 租金
    [date] [date] NULL ,                             -- 发布时间
    [telephone] [varchar] (50) COLLATE Chinese_PRC_CI_AS NULL ,  -- 联系电话
    [lxr] [varchar] (50) COLLATE Chinese_PRC_CI_AS NULL      -- 联系人
) ON [PRIMARY]
GO

-- 街道表
CREATE TABLE [dbo].[TBL_JD] (
    [jdid] [int] IDENTITY (6, 1) NOT NULL ,          -- 街道编号
    [jd] [varchar] (50) COLLATE Chinese_PRC_CI_AS NULL ,     -- 街道名称
    [qxid] [int] NOT NULL                            -- 所属区县编号
) ON [PRIMARY]
GO
```

```
-- 区县表
CREATE TABLE [dbo].[TBL_QX] (
    [qxid] [int] NOT NULL ,                            -- 区县编号
    [qx] [varchar] (50) COLLATE Chinese_PRC_CI_AS NULL    -- 区县名称
) ON [PRIMARY]
GO

-- 用户表
CREATE TABLE [dbo].[TBL_USER] (
    [uid] [int] IDENTITY (1, 1) NOT NULL ,             -- 用户编号
    [uname] [varchar] (50) COLLATE Chinese_PRC_CI_AS NOT NULL ,  -- 用户名
    [upass] [varchar] (50) COLLATE Chinese_PRC_CI_AS NOT NULL    -- 密码
) ON [PRIMARY]
GO
```

要完成 Hibernate 的"对象—关系"映射需要做如下的准备工作。

● 需要的 jar 包

Hibernate 的官方主页是 www.hibernate.org，我们需要的 jar 包都可以从官方网站上下载得到。在 MyEclipse 中，MyEclipse 插件已经集成了对 Hibernate 开发的支持，所以我们不必自己下载 Hibernate 依赖的 jar 包。

在 MyEclipse 中，在项目节点上单击右键，从弹出的快捷菜单上选择"MyEclipse"→"Add Hibernate Capabilities"选项，如图 4-4 所示。

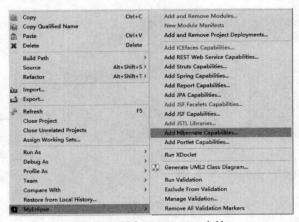

图 4-4　添加 Hibernate 支持

在弹出的"Add Hibernate Capabilities"对话框中，选择"Hibernate 4.1"单选项。如果想查看 MyEclipse 将自动为我们添加哪些包，可以单击"View and edit libraries…"按钮查看。如图 4-5 所示。

第 4 章　使用 Hibernate 完成对象持久化

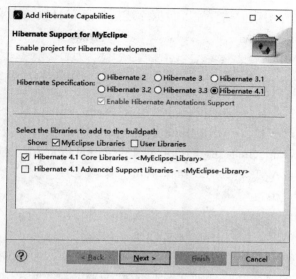

图 4-5　单看 Hibernate 库文件

在一般项目开发中，都是通过 MyEclipse 工具添加这些 Hibernate 的 jar 包。当然也可以从 http://www.hibernate.org 下载。这些 jar 包中，hibernate4.jar 是 Hibernate 核心包；asm 相关的包用来支持字节码的动态生成，是必需的；common-* 的包是 Apache Common 项目提供的，包含集合工具类、日志接口等，Hibernate 实现过程中使用了这些开源工具。Hibernate4.1 核心包如图 4-6 所示。

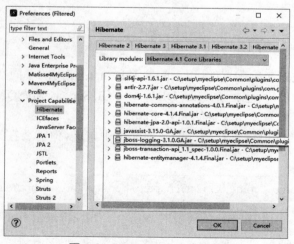

图 4-6　Hibernate 3.2 库文件

在"Specify Hibernate database connection details"对话框中，取消复选框的选中状态，先不配置数据库连接。

● Hibernate 配置文件

另一项准备工作就是在项目中添加 Hibernate 配置文件，默认文件名为"hibernate.cfg.xml"。在使用 MyEclipse 工具给项目添加 Hibernate 支持的时候会自动给我们添加 Hibernate

的配置文件。在配置文件中我们需要配置数据库连接信息和 Hibernate 的参数，hibernate.cfg.xml 如示例代码 4-2 所示。

```xml
示例代码 4-2：hibernate.cfg.xml 源代码片段
<?xml version='1.0' encoding='UTF-8'?>
<!DOCTYPE hibernate-configuration PUBLIC
        "-//Hibernate/Hibernate Configuration DTD 3.0//EN"
        "http://www.hibernate.org/dtd/hibernate-configuration-3.0.dtd">
<!-- Generated by MyEclipse Hibernate Tools. -->
<hibernate-configuration>

    <session-factory>
        <property name="dialect">org.hibernate.dialect.SQLServerDialect</property>
        <property name="connection.url">jdbc:sqlserver://localhost:1433</property>
        <property name="connection.username">sa</property>
        <property name="connection.password">123456</property>
        <property name="connection.driver_class">com.microsoft.sqlserver.jdbc.SQLServerDriver</property>
        <property name="myeclipse.connection.profile">zf_db</property>

    </session-factory>

</hibernate-configuration>
```

其中，dialect 参数是必须配置的，用于配置 Hibernate 使用的不同数据库类型。Hibernate 支持几乎所有的主流数据库，包括 Oracle、DB2、MSSQL Server、MySQL 等。show_sql 参数为 true，则程序运行时在控制台输出执行的 SQL 语句。

数据库表和实体的映射信息在另外的映射文件中定义，但需要在配置文件中声明，具体如下：

```xml
<mapping resource="com/xtgj/j2ee/chapter04/entity/TblUser.hbm.xml" />
```

上面的代码表示初始化 Hibernate 环境时，将装载 TblUser.hbm.xml 映射信息。
● 实体类和映射文件
我们要使用 Hibernate 实现添加用户的功能，首先要定义用户实体类，把之前定义的用户实体类做两个修改就可以了：实现 implements java.io.Serializable 接口；添加默认构造方法。如示例代码 4-3 所示。

示例代码 4-3：实体类 TblUser.java 源代码片段

```java
package com.xtgj.j2ee.chapter04.entity;

import java.util.HashSet;
import java.util.Set;

public class TblUser implements java.io.Serializable {
    private Integer uid;
    private String uname;
    private String upass;
    private Set tblFwxxes = new HashSet(0);

    public TblUser() {
    }

    public TblUser(String uname, String upass) {
        this.uname = uname;
        this.upass = upass;
    }

    public TblUser(String uname, String upass, Set tblFwxxes) {
        this.uname = uname;
        this.upass = upass;
        this.tblFwxxes = tblFwxxes;
    }

    public Integer getUid() {
        return uid;
    }

    public void setUid(Integer uid) {
        this.uid = uid;
    }

    public String getUname() {
        return uname;
    }
```

```java
    public void setUname(String uname) {
        this.uname = uname;
    }

    public String getUpass() {
        return upass;
    }

    public void setUpass(String upass) {
        this.upass = upass;
    }

    public Set getTblFwxxes() {
        return tblFwxxes;
    }

    public void setTblFwxxes(Set tblFwxxes) {
        this.tblFwxxes = tblFwxxes;
    }
}
```

然后还需要告诉 Hibernate，实体类 User 映射到数据库哪个表，以及哪个属性对应到数据库表的哪个字段，这些在映射文件中配置。TblUser.hbm.xml 文件配置如示例代码 4-4 所示。

示例代码 4-4：映射文件 TblUser.hbm.xml 源代码片段

```xml
<?xml version="1.0"?>
<!DOCTYPE hibernate-mapping PUBLIC "-//Hibernate/Hibernate Mapping DTD 3.0//EN"
    "http://hibernate.sourceforge.net/hibernate-mapping-3.0.dtd">
<!--
    Mapping file autogenerated by MyEclipse - Hibernate Tools
-->
<hibernate-mapping>
    <class name="com.xtgj.j2ee.chapter04.entity.TblUser" table="TBL_USER" schema="dbo"
        catalog="zf">
        <id name="uid" type="java.lang.Integer">
            <column name="uid" />
```

```xml
            <generator class="native" />
        </id>
        <property name="uname" type="java.lang.String">
        <property name="upass" type="java.lang.String">
            <column name="upass" length="50" not-null="true" />
        </property>
        <!--<set name="tblFwxxes" inverse="true">
            <key>
                <column name="uid" />
            </key>
            <one-to-many class="com.xtgj.j2ee.chapter04.entity.TblFwxx" />
        </set>-->
    </class>
</hibernate-mapping>
```

示例代码 4-4 展示了用户实体 TBL_USER 映射的配置。在映射文件中，每个 <class> 节点配置一个实体类的映射信息，<class> 节点的 name 属性对应实体类的名字，table 属性对应数据库表的名字。在 <class> 节点下，有一个必需的 <id> 节点，用于定义实体的标识属性（对应数据库表的主键）。<id> 节点的 name 属性对应实体类的属性，type 为对应的 Java 类型。<id> 节点下有两个子节点，<column> 用于通过其 name 属性指定对应的数据库表的主键，<generator> 节点用于指定主键的生成策略，常用的值有 native 和 assigned，native 表示由数据库生成主键的值，assigned 表示在添加新记录到数据前由程序设定主键的值。<class> 节点下除了 <id> 节点，还包括 <property> 子节点。<property> 节点与 <id> 节点类似，只是不能包括 <generator> 子节点。每个 <property> 节点指定一对属性和字段的对应关系。

4.3.2 使用 Hibernate 实现用户添加

我们提出这样的问题：使用 Hibernate 编写代码，实现添加新用户到数据库的功能。

分析：使用 Hibernate 执行持久化操作（数据库增删查改）需要 7 个步骤，我们可以类比 JDBC 理解。见表 4-1。

那么，我们就产生了如下的疑问：使用 Hibernate 居然需要 7 个步骤，那不是比使用 JDBC 还麻烦？回答是否定的，因为实际上在代码开发的时候，我们并不需要关注那么多环节，具体的方法将在 4.6 节介绍。

表 4-1 Hibernate 执行持久化操作的步骤

步骤	操作	代码	与 JDBC 类比
1	读取并解析配置文件	Configuration conf=new Configuration.configure()	相当于使用 DataSource 获取连接前读取 DataSource 的配置文件

续表

步骤	操作	代码	与 JDBC 类比
2	读取并解析映射信息，创建 SessionFactory	SessionFactory sf = conf.bulidSessionFactory()	相当于创建 DataSource 对象
3	打开 session	Session session = sf.openSession()	相当于 JDBC 获得连接
4	开始一个事物（增删改操作必须，查询操作可选）	Transaction tx = seesion.beginTransaction()	开始事务
5	持久化操作	session.save(user)	持久化操作
6	提交事务	tx.commit()	提交事务
7	关闭 session	session.close()	关闭连接

代码如示例代码 4-5 所示。

示例代码 4-5：实现添加新用户到数据库

```java
package com.xtgj.j2ee.chapter04.dao;

import org.hibernate.Session;
import org.hibernate.SessionFactory;
import org.hibernate.Transaction;
import org.hibernate.cfg.Configuration;

import com.xtgj.j2ee.chapter04.entity.TblUser;

public class HibernateTest1 {
    public static void main(String[] args) throws Exception{
        // 1. 读取配置文件
        Configuration conf = new Configuration().configure();
        // 2. 创建 SessionFactory
        SessionFactory sf = conf.buildSessionFactory();
        // 3. 打开 session
        Session session = sf.openSession();
        Transaction tx = null;
        try {
            // 4. 开始一个事物
            tx = session.beginTransaction();
            // 5. 持久化操作
            TblUser user = new TblUser();
```

```
                user.setUname("Hibernate user");
                user.setUpass("password");
                session.save(user);
                // 6. 提交事物
                tx.commit();
            } catch (Exception e) {
                e.printStackTrace();
                if (tx != null)
                        tx.rollback();
                throw e;
            } finally {
                // 7. 关闭 Session
                session.close();
            }
        }
    }
```

与 JDBC 类似，持久化操作要放在 try{} 语句块中，如果发生异常则将事务回滚。关闭 session 的语句放在 finally{} 语句块中。示例代码 4-5 的执行结果如图 4-7 所示。

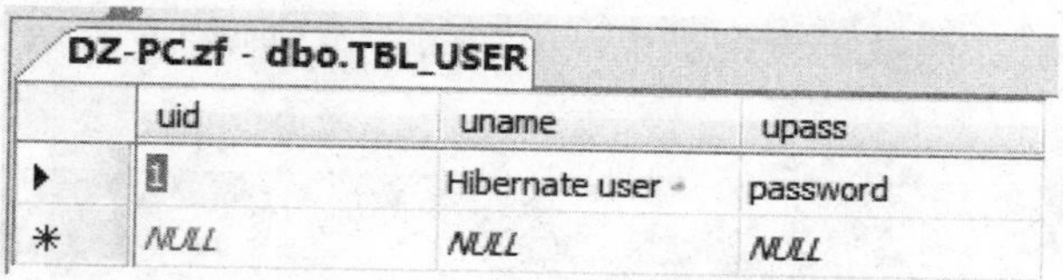

图 4-7　使用 Hibernate 添加用户

至此，我们就完成了使用 Hibernate 添加数据的任务。

示例代码 4-5 中涉及了几个 Hibernate 的类，包括 Configuration、SessionFactory、Session 和 Transaction，对比 DataSource 和 JDBC 的概念很容易理解。图 4-8 展示了几个类间的关系及 Hibernate 执行的过程。

回想一下，使用 Hibernate 需要"3 个准备，7 个步骤"，各是什么呢？试一下，能不能默写出示例代码 4-5 的 7 个步骤。这段代码是使用 Hibernate 的经典"套路"！

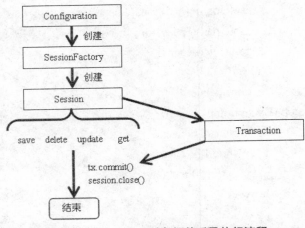

图 4-8　Hibernate 对象间关系及执行流程

4.4　使用 Hibernate 实现数据的删改

上面演示了如何使用 Hibernate 添加数据。下面我们学习一下如何使用 Hibernate 加载、修改和删除数据。

加载是根据主键将一条数据查询出来。下面的示例代码 4-6 展示了 Hibernate 加载的全过程。

示例代码 4-6：Hibernate 加载

```
package com.xtgj.j2ee.chapter04.dao;

import org.hibernate.Session;
import org.hibernate.SessionFactory;
import org.hibernate.cfg.Configuration;

import com.xtgj.j2ee.chapter04.entity.TblUser;

public class HibernateTest2 {
    private SessionFactory sf = new Configuration().configure()
            .buildSessionFactory();

    public static void main(String[] args) throws Exception {
        new HibernateTest2().testLoad();
    }
```

```java
    public TblUser get(Integer id) {
        Session session = sf.openSession();
        TblUser user = (TblUser) session.get(TblUser.class, id);
        session.close();
        return user;
    }
    public void testLoad() throws Exception {

        TblUser user = this.get(1);
        System.out.println(user.getUname() + "," + user.getUpass());
    }

}
```

如示例代码 4-6 所示，使用 Hibernate 加载数据只需要一行代码，不再需要繁琐地从 ResultSet 中取数据封装到实体中。

下面我们看修改操作是怎么完成的。如示例代码 4-7 所示。

示例代码 4-7：Hibernate 修改操作

```java
package com.xtgj.j2ee.chapter04.dao;

import org.hibernate.Session;
import org.hibernate.SessionFactory;
import org.hibernate.Transaction;
import org.hibernate.cfg.Configuration;

import com.xtgj.j2ee.chapter04.entity.TblUser;

public class HibernateTest3 {

    private SessionFactory sf = new Configuration().configure()
            .buildSessionFactory();

    public static void main(String[] args) throws Exception {
        new HibernateTest3().testUpdate();
    }

    public TblUser get(Integer id) {
```

```java
        Session session = sf.openSession();
        TblUser user = (TblUser) session.get(TblUser.class, id);
        session.close();
        return user;
    }

    public void testUpdate() throws Exception {

        Session session = sf.openSession();
        Transaction tx = null;
        try {
            tx = session.beginTransaction(); // 增删改操作一定能够要开启事务
            TblUser user = this.get(1); // 先加载,再修改
            user.setUname("new Name");
            session.update(user); // 执行修改的代码
            tx.commit();
        } catch (Exception e) {
            if (tx != null)
                tx.rollback();
            throw e;
        } finally {
            session.close();
        }
    }
}
```

只需要简单调用 session 的 update 方法即可。需要注意的是,增删改操作一定要在事务环境中完成。

那么删除操作又是怎么完成的呢?请看如示例代码 4-8 所示的代码。

示例代码 4-8:Hibernate 删除操作

```java
package com.xtgj.j2ee.chapter04.dao;

import org.hibernate.Session;
import org.hibernate.SessionFactory;
import org.hibernate.Transaction;
import org.hibernate.cfg.Configuration;

import com.xtgj.j2ee.chapter04.entity.TblUser;
```

```java
public class HibernateTest4 {

    private SessionFactory sf = new Configuration().configure()
                    .buildSessionFactory();

    public static void main(String[] args) throws Exception {

        new HibernateTest4().testDel();
    }
    public TblUser get(Integer id) {
        Session session = sf.openSession();
        TblUser user = (TblUser) session.get(TblUser.class, id);
        session.close();
        return user;
    }
    public void testDel() throws Exception {
        Session session = sf.openSession();
        Transaction tx = null;
        try {
            tx = session.beginTransaction();
            TblUser user = this.get(1); // 先加载,再删除
            session.delete(user); // 执行删除的代码
            tx.commit();
        } catch (Exception e) {
            if (tx != null)
                tx.rollback();
            throw e;
        } finally {
            session.close();
        }
    }
}
```

与修改类似,删除时也需要先加载数据。在使用 Hibernate 编写持久化代码时,不需要再有数据库表、字段等概念。根据面向对象的语义,删除方法的参数是将要删除的对象,而不是主键。以面向对象的思维编写代码是 Hibernate 持久化操作接口设计的一个理念。

4.5 使用工具简化 Hibernate 开发

现在我们已经学会了如何使用 Hibernate 完成持久化操作,但"3 个准备,7 个步骤"是不是太啰嗦了啊!别急,实际开发中有很多办法可以简化这一过程。我们可以从两个角度减轻负担:一是充分利用工具,二是精巧的设计。

我们首先看一下如何使用 MyEclipse 简化 Hibernate 的开发。

在 4.4 节,我们使用 MyEclipse 工具给项目添加 Hibernate 支持,自动添加了 Hibernate 配置文件。单击 MyEclipse Database Explorer 按钮,在弹出的 Database Driver 对话框中,Driver name 可以随意。URL 一定要填写正确,单击"下一步"按钮,如图 4-9 所示。

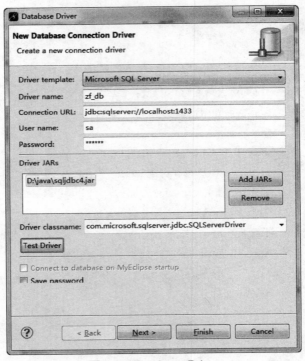

图 4-9 Database Driver

在 Database Driver 对话框中,选择要使用的数据库,单击"完成"按钮。MyEclipse 就会自动在 Hibernate 配置文件中配置好数据库相关参数,如图 4-10 所示。

那么映射文件呢?例如,TblUser.hbm.xml,是不是也有工具辅助完成呢?是的,映射文件完全可以使用工具自动生成。

在 MEclipse 工作区右上角选择进入 MyEclipse Database Explorer 透视图(也可以通过菜单"窗口"→"透视图"→"MyEclipse Database Explorer"进入),如图 4-11 所示。

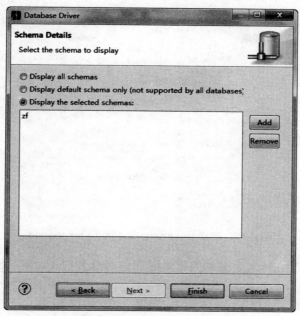

图 4-10　Database Driver

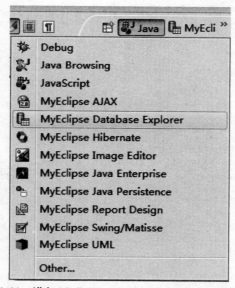

图 4-11　进入 MyEclipse Database Explorer 透视图

在 DB Browser（数据库浏览器）视图中，逐次展看节点，用鼠标右键选择 TBL_USER 表，在弹出的菜单中选择"Hibernate Reverse Engineering"（Hibernate 反向工程）选项，如图 4-12 所示。

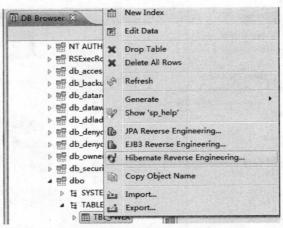

图 4-12 启动 Hibernate Reverse Engineering

在弹出的对话框中选择生成实体类和映射文件的包名（Java Package）。选中"Create PO-JO<>DB table mapping information"复选框，表示要生成映射文件；选中"Java Data Object…"复选框表示要生成实体类。然后单击"下一步"按钮，如图 4-13 所示。

图 4-13 配置生成项

在"Config type mapping details"对话框中，ID Generator 选"native"，继续单击"下一步"按钮。在"Config reverse engineering details"中保持所有复选框未选中（关联的内容下一章介绍），单击"完成"按钮，开始生成代码。完成后，返回 Java 透视图，发现已经生成了实体类和映射文件，如图 4-14 所示。

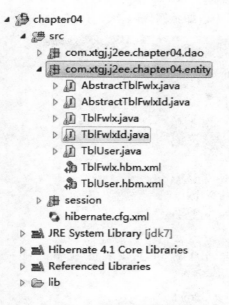

图 4-14 生成的实体类和映射文件

现在,我们可以不用敲一行代码完成"3 个准备",是不是比 JDBC 简化许多。

4.6 在项目中使用 Hibernate

前面介绍了使用 MyEclipse 简化"3 个准备",但"7 个步骤"应该如何简化?我们曾经使用 BaseJdbcDAO 简化 JDBC 的编码,现在采用类似的方法简化 Hibernate 的编码。在给项目添加 Hibernate 支持的过程中还生成了一个 HibernateSessionFactory 工具类,我们也要利用一下。

使用 Hibernate 重新实现 FwxxDAO 接口,如示例代码 4-9 所示。

示例代码 4-9:FwxxDAO 接口

```
package com.xtgj.j2ee.chapter04.basedao;
import java.util.List;
import com.xtgj.j2ee.chapter04.entity.FWXX;
public interface FwxxDAO {
    public void add(FWXX item);
    public void del(int fwid);
    public void update(FWXX item);
    public List search(FWXX condition);
}
```

观察示例代码 4-7 和示例代码 4-8,大部分代码是"套路"代码,我们现在就把这些"套路"

代码提取到 BaseHibernateDAO 中。如示例代码 4-10 所示。

示例代码 4-10：BaseHibernateDAO.java 源代码

```java
package com.xtgj.j2ee.chapter04.basedao;

import java.io.Serializable;
import java.util.List;

import org.hibernate.Session;
import org.hibernate.Transaction;
import org.hibernate.criterion.Example;

import com.xtgj.j2ee.chapter04.session.HibernateSessionFactory;

public abstract class BaseHibernateDAO {

    private Session session = null;

    public void setSession(Session session) {
        this.session = session;
    }

    protected Session getSession() {
        this.session = HibernateSessionFactory.getSession();
        return this.session;
    }

    protected void closeSession() {
        this.session = null;
        HibernateSessionFactory.closeSession();
    }

    protected void add(Object item) {

        Transaction tx = null;
        Session session = getSession();
        try {
            tx = session.beginTransaction();
```

```java
                session.save(item);
                tx.commit();
        } catch (Exception e) {
                e.printStackTrace();
        } finally {
                closeSession();
        }
}

protected Object get(Class clz, Serializable id) {

        Object item = null;
        Session session = getSession();
        try {
                item = session.get(clz, id);
        } catch (Exception e) {

                e.printStackTrace();
        } finally {
                closeSession();
        }
        return item;
}

protected void delete(Class clz, Serializable id) {
        Transaction tx = null;
        Session session = getSession();
        try {
                tx = session.beginTransaction();
                getSession().delete(this.get(clz, id));
                tx.commit();
        } catch (Exception e) {
                if (null != tx) {
                        tx.rollback();
                }
                e.printStackTrace();
        } finally {
                closeSession();
```

```
            }
        }

        protected void update(Object item) {
            Transaction tx = null;
            Session session = getSession();
            try {
                tx = session.beginTransaction();
                session.update(item);
                tx.commit();
            } catch (Exception e) {
                if (null != tx) {
                    tx.rollback();
                }
                e.printStackTrace();
            } finally {
                closeSession();
            }
        }

        protected List search(Class clazz, Object condition) {
            try {
                List results = getSession().createCriteria(clazz).add(
                        Example.create(condition)).list();
                return results;
            } catch (RuntimeException re) {
                throw re;
            } finally {
                closeSession();
            }
        }
    }
```

这里只提供了简单的查询方法，Hibernate 查询我们将在第 6 章学习。需要注意的是，get 和 delete 的传入参数类型为 Class 和 Serializable。Class 表示"类"，可以通过"类名 .class"得到其实例，如"User.class"。Serializable 是一个接口，不包含任何需要实现的方法，只是表示实现这个接口的类可以序列化。

现在我们就继承 BaseHibernateDAO 类并实现 FwxxDAO 接口，完成房屋信息管理的业务逻辑。如示例代码 4-11 所示。

示例代码 4-11：实现 FwxxDAO 接口

```
package com.xtgj.j2ee.chapter04.dao;

// 省略 import

public class FwxxDAOhibImpl extends BaseHibernateDAO implements FwxxDAO {

    public FWXX load(int fwid) {
        return (FWXX) super.get(FWXX.class, fwid);
    }

    public void add(FWXX fwxx) {
        super.add(fwxx);
    }

    public void del(int fwid) {
        super.delete(FWXX.class, fwid);
    }

    public void update(FWXX fwxx) {
        super.update(fwxx);
    }

    @SuppressWarnings("unchecked")
    public List search(FWXX condition) {
        List ret = super.search(FWXX.class, condition);
        return ret;
    }
}
```

现在 DAO 层的代码是不是无比清爽，这正是使用 Hibernate 的目的所在。看到 Hibernate 的魔力了吧！下面我们将继续学习 Hibernate，第 5 章将介绍 Hibernate 如何映射实体间的关联关系，第 6 章将学习 Hibernate 查询。

4.7 小结

- Hibernate 是一个基于 JDBC 的持久化解决方案，是一个优秀的"对象—关系"映射框架。
- ✓ 使用 Hibernate 前要做好以下 3 个准备：
  - ◆ 添加需要的 jar 包；
  - ◆ 配置 Hibernate 配置文件；
  - ◆ 创建实体类和映射文件。
- ✓ 使用 Hibernate 完成持久化操作需要以下 7 个步骤：
  - ◆ 读取并解析配置文件；
  - ◆ 读取并解析映射文件，创建 SessionFactory；
  - ◆ 打开 session；
  - ◆ 开启事务（查询操作不需要）；
  - ◆ 执行持久化方法；
  - ◆ 提交事务；
  - ◆ 关闭 session。
- ✓ 使用 MyEclipse 提供的 Hibernate 工具可以简化"3 个准备"。
- ✓ 使用提取重复代码到基类的技巧可以简化"7 个步骤"。

4.8 英语角

session	会话
transient	瞬时
persistent	持久
relational	关系
mapping	映射

4.9 作业

1. 利用本章中的 BaseHibernateDAO 分别完成用户的增删改查。
2. 利用本章中的 BaseHibernateDAO 分别完成房屋信息的增删改查。

4.10 思考题

Hibernate 是如何实现对数据的增删改查操作的？

4.11 学员回顾内容

1. 使用 Hibernate 的基本步骤。
2. Hibernate 的基本原理。

第 5 章　Hibernate 的关联映射

学习目标

- ◆ 掌握单向多对一关联。
- ◆ 掌握单向一对多关联。
- ◆ 掌握双向一对多关联。
- ◆ 掌握多对多关联。

课前准备

通过互联网了解 Hibernate 的关联。

本章简介

前面一章我们了解使用 Hibernate 完成持久化操作的 7 个步骤,以及使用 MyEclipse 提供的 Hibernate 工具可以简化"3 个准备",使用提取重复代码到基类的技巧可以简化"7 个步骤"。初步了解 Hibernate 后,它们之间的关联又是怎样的呢?让我们带着好奇来学习本章。

5.1　关联

5.1.1　面向对象领域的关联关系

我们首先回顾一下面向对象领域内对象和对象间的关联关系。以租房系统中的区县和街道为例,区县和街道之间的关系是典型的一对多关系,一个区县下管辖多个街道,多个街道同属于一个区县。

关联是有方向的,如图 5-1 所示就表现了街道到区县的单向多对一关系。

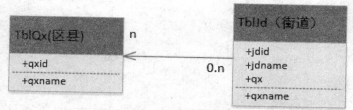

图 5-1 街道到区县的单向多对一关联

因为在街道类 TblJd 中,有一个 qx 属性关联到区县类 TblQx,而区县类并没有关联到街道类。对实体类来说,存在这样的关联给我们编码带来了很大方便。回想一下,当我们通过 session.get(FWXX.class,id) 加载到房屋信息的实体对象后,对象中只有街道的 id,发布信息用户的 id,如果想在页面上显示街道的名称,或发布信息用户的名称,还要再次进行查询。像图 5-1 中这种情况,街道(TblJd)对象里不是保存区县(TblQx)的 id,而是保存了区县对象,我们就可以通过"TblJd.getQx().getQxname();"直接取得区县的名称。当然,想要 Hibernate 加载对象时自动加载关联的对象的前提是我们在对应的映射文件(如 TblJd.hbm.xml)中做好了配置(使用 Hibernate,大多时候我们是无须关心 SQL 语句的)。我们马上就来学习是怎么配置的。

与单向多对一类似,也存在着单向的一对多关联,如图 5-2 所示。例如:一个区县类中定义了关联到街道属性。这时,由于是一个对象关联到了多个,所以属性的类型一定是集合类型。这样就可以通过如下的代码在"一"的一方访问与之关联的"多"的一方的数据。

```
示例代码 5-1:单向关系的一对多关系中实体类定义
    public void findJD(int qxid){
        TblQx qx = (TblQx)super.get(TblQx.class, qxid);
        TblJd jd= (TblJd)(qx.getJds().iterator().next());
        System.out.println(jd.getJd());
    }
    public static void main(String[] args) {
        new RelationTest1().findJD(1);
    }
```

当然,也可以遍历处理集合中的每个数据。

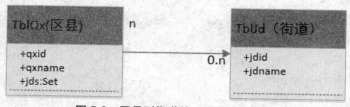

图 5-2 区县到街道的一对多单向关联

将双方都关联到对方的时候,就称为存在双向一对多关联(称为双向多对一关联亦可),如图 5-3 所示。

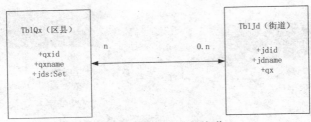

图 5-3　双向一对多关联

5.1.2　关联关系的映射配置

与对象关系中的关联关系类似，数据库的表与表之间也存在着关联关系，主要以外键的形式体现。图 5-4 所示的是区县表和街道表间的关联关系。

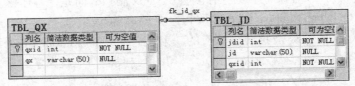

图 5-4　数据库表间的一对多关系

我们如何使用 Hibernate 在两者之间映射呢？

● 单向多对一关联的映射

我们首先考虑单向多对一的情况。实体类的定义如示例代码 5-2 所示。通过 jd 类的 qx 属性体现了两个类之间的关联。jd 属于多的一方，所以这是一种典型的单向多对一关联，代码如示例代码 5-2 所示。

示例代码 5-2：单向多对一关系中的实体类定义

```
//区县实体
public class TblQx implements java.io.Serializable {
    private int qxid;
    private String qx;

    public String getQx() {
        return qx;
    }

    public void setQx(String qx) {
        this.qx = qx;
    }

    public int getQxid() {
        return qxid;
```

```java
    }
    public void setQxid(int qxid) {
        this.qxid = qxid;
    }
}
// 街道实体
public class TblJd implements Serializable {
    private Integer jdid=1;
    private String jd;
    private TblQx qx;

    public TblQx getQx() {
        return qx;
    }

    public void setQx(TblQx qx) {
        this.qx = qx;
    }

    public String getJd() {
        return jd;
    }

    public void setJd(String jd) {
        this.jd = jd;
    }

    public int getJdid() {
        return jdid;
    }

    public void setJdid(int jdid) {
        this.jdid = jdid;
    }
```

然后看看两个类的映射文件是怎么配置的，如示例代码 5-3 所示。

示例代码 5-3：单向多对一映射文件配置

```xml
<!-- 区县实体映射信息 -->
<hibernate-mapping >
    <class name="com.xtgj.j2ee.chapter05.entity.TblQx" table="TBL_QX" schema="dbo" catalog="zf">
        <id name="qxid" column="qxid" type="integer"/>
        <property name="qx" column="qx" type="string" not-null="false"/>
    </class>
</hibernate-mapping><!-- 街道实体映射信息 -->
<hibernate-mapping>
    <class name="com.xtgj.j2ee.chapter05.entity.TblJd" table="TBL_JD" schema="dbo" catalog="zf" >
        <id name="jdid" column="jdid" type="java.lang.Integer"/>
        <property name="jd" column="jd" type="java.lang.String" not-null="false"/>
        <many-to-one name="qx" class="com.xtgj.j2ee.chapter05.entity.TblQx" column="qxid"></many-to-one>
    </class>
</hibernate-mapping>
```

大部分配置代码正如上一章所学的，只有 <many-to-one name="qx" class=" TblQx " column="qxid"/> 一句是新加的。其中 name 属性对应 TblJd 类的属性名 (qx)；column 属性定义了映射到表 TBL_JD 的字段名，将使用这个字段作为外键去和"一"的一方主键关联；class 属性为关联到"一"的一方的类别。很简单吧！现在我们就编写程序代码，验证配置的多对一关系是否生效，如示例代码 5-4 所示。BaseHibernateDAO 代码见第 4 章示例代码 4-10。

示例代码 5-4：单向多对一测试

```java
package com.xtgj.j2ee.chapter05.dao;

import com.xtgj.j2ee.chapter05.basedao.BaseHibernateDAO;
import com.xtgj.j2ee.chapter05.entity.TblJd;

public class ManyToOneTest extends BaseHibernateDAO {
    public static void main(String[] args) {
        new ManyToOneTest().testManyToOne();
    }

    public void testManyToOne() {
        TblJd JD = (TblJd) super.get(TblJd.class, 1);
        String qxname = JD.getQx().getQx();
```

第 5 章 Hibernate 的关联映射

```
            System.out.println(qxname + "," + JD.getJd());
    }
}
```

只是加载了 jd 对象，jd 可以访问关联的 TblQx 对象的属性，正如我们期望的，运行结果如图 5-5 所示，可以看到 Hibernate 自动执行了两条 select 语句。

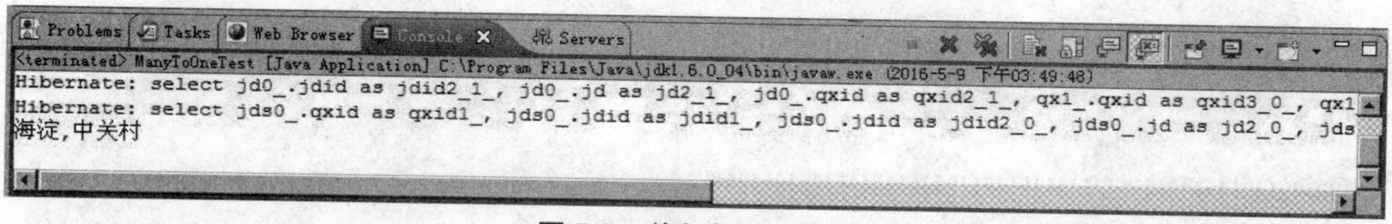

```
<terminated> ManyToOneTest [Java Application] C:\Program Files\Java\jdk1.6.0_04\bin\javaw.exe (2016-5-9 下午03:49:48)
Hibernate: select jd0_.jdid as jdid2_1_, jd0_.jd as jd2_1_, jd0_.qxid as qxid2_1_, qx1_.qxid as qxid3_0_, qx1
Hibernate: select jds0_.qxid as qxid1_, jds0_.jdid as jdid1_, jds0_.jdid as jdid2_0_, jds0_.jd as jd2_0_, jds
海淀,中关村
```

图 5-5　单向多对一示例

● 单向一对多关联的映射

现在我们在示例代码 5-1 的 TblQx 类的定义中加上 Set 类型的 jds 属性。最终 TblQx 类的定义如示例代码 5-5 所示。

示例代码 5-5：单向一对多关系中的实体类定义

```java
//区县实体
package com.xtgj.j2ee.chapter05.dao;

import java.io.Serializable;
import java.util.HashSet;

import org.hibernate.mapping.Set;


public class TblQx implements Serializable {
    private int qxid=0;
    private String qx="";
    private HashSet jds = new HashSet();
    public HashSet getJds() {
            return jds;
    }
    public void setJds(HashSet jds) {
            this.jds = jds;
    }
    public int getQxid() {
            return qxid;
    }
```

```java
    public void setQxid(int qxid) {
        this.qxid = qxid;
    }
    public String getQx() {
        return qx;
    }
    public void setQx(String qx) {
        this.qx = qx;
    }



}
// 街道实体
public class TblJd implements java.io.Serializable {
    private int jdid;
    private String jd;
    // 省略 getter/setter
}
```

然后在示例代码 5-2 的 qx 类映射信息中加上一对多的配置。最终 TblQx 的配置信息如示例代码 5-6 所示。

示例代码 5-6：TblQx 的配置信息

```xml
<?xml version="1.0"?>
<!DOCTYPE hibernate-mapping PUBLIC "-//Hibernate/Hibernate Mapping DTD 3.0//EN"
    "http://hibernate.sourceforge.net/hibernate-mapping-3.0.dtd">

<hibernate-mapping >
    <class name="com.xtgj.j2ee.chapter05.entity.TblQx" table="TBL_QX" schema="dbo" catalog="zf">
        <id name="qxid" column="qxid" type="integer"/>
        <property name="qx" column="qx" type="string" not-null="false"/>
        <set name="jds" inverse="true" lazy="false" cascade="all">
            <key>
                <column name="qxid" />
            </key>
            <one-to-many class="com.xtgj.j2ee.chapter05.entity.TblJd" />
        </set>
```

```
    </class>

</hibernate-mapping>
```

比多对一(many-to-one)的配置稍微麻烦一点,需要创建一个 <set> 节点,其 name 属性对应到 qx 类的 jds 属性。<set> 节点下需要建立两个子节点:一个 <key> 节点,通过其 column 属性指定"多"的一方的外键字段名(注意:这里是 TBL_JD 表的字段名);一个 <one-to-many> 节点,通过其 class 属性指定关联到"多"的一方的类别。只需要配置这 3 个信息即可。

至此,我们就完成了添加单向一对多配置的工作。现在我们测试一下配置的效果,如示例代码 5-7 所示。

示例代码 5-7:测试单向一对多

```java
package com.xtgj.j2ee.chapter05.dao;

import java.util.Iterator;
import com.xtgj.j2ee.chapter05.basedao.BaseHibernateDAO;
import com.xtgj.j2ee.chapter05.entity.TblJd;
import com.xtgj.j2ee.chapter05.entity.TblQx;

public class OneToManyTest extends BaseHibernateDAO {
    public static void main(String[] args) {
        new OneToManyTest().testOneToMany();
    }

    public void testOneToMany() {
        TblQx QX = (TblQx) super.get(TblQx.class, 1);
        // 输出
        System.out.println(" 区县 "+QX.getQx()+" 的街道:");// 输出区县名
        Iterator it = QX.getJds().iterator();// 迭代输出区县下属的街道
        while (it.hasNext()) {
            TblJd jd = (TblJd) it.next();
            System.out.print(jd.getJd() + ",");
        }
    }
}
```

只通过加载 qx 数据就自动加载了区县下属的街道数据,运行效果如图 5-6 所示。

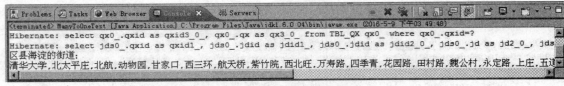

图 5-6 单向一对多示例

- 双向一对多关联的映射

单向一对多和单向多对一可以分别配置使用。如果同时配置了两者,就成了双向一对多关联。其实在前面我们就已经完成了双向一对多关联的配置。

配置关联时,我们需要考虑两点:第一,程序执行效率,第二,实际业务的需要。比如在前面的例子中,加载街道的同时获得对应区县的信息,这给编码带来了极大的便利。但是,我们并不是很需要通过区县对象直接得到下属的所有街道的数据,加载区县数据时就自动加载街道信息也会带来性能上的损失。这时,我们只需要配置单向的多对一关联就好了。

5.1.3 通过程序维护关系

我们提出这样的问题:在前面配置好双向一对多后,考虑如何实现下列功能。
(1) 增加一个区县:"山南区"。
(2) 增加山南区下属的 3 个街道:"和平路"、"八一路"和"五四大道"。
(3) 删除"五四大道"和"八一路"。
(4) 将"和平路"划分到海淀区。
(5) 删除山南区。

分析:上述逻辑功能都是处理一对多关系时常见的业务需求。上一章我们学到,通过 session.save(Object) 即可将对象添加到数据库,通过 session.update(Object) 即可更新数据,现在我们还是这样做。这时,我们需要规定区县数据的主键生成策略,即配置区县实体的 id 属性,区县实体的映射文件修改如下:

```xml
<?xml version="1.0"?>
<!DOCTYPE hibernate-mapping PUBLIC "-//Hibernate/Hibernate Mapping DTD 3.0//EN"
"http://hibernate.sourceforge.net/hibernate-mapping-3.0.dtd">

<hibernate-mapping >
    <class name="com.xtgj.j2ee.chapter05.entity.TblQx" table="TBL_QX" schema="dbo" catalog="zf">
        <id name="qxid" column="qxid" type="integer">
            <generator class="identity"></generator>
        </id>
        <property name="qx" column="qx" type="string" not-null="false"/>
        <set name="jds" inverse="true" lazy="false" cascade="all">
```

```xml
            <key>
                <column name="qxid" />
            </key>
            <one-to-many class="com.xtgj.j2ee.chapter05.entity.TblJd" />
        </set>
    </class>

</hibernate-mapping>
```

以添加"山南区"为例,由于我们的示例类继承了 BaseHibernateDAO 类,所以只需要调用父类的 add 方法即可。如示例代码 5-8 所示。

示例代码 5-8:添加"山南区"

```java
package com.xtgj.j2ee.chapter05.dao;

import com.xtgj.j2ee.chapter05.basedao.BaseHibernateDAO;
import com.xtgj.j2ee.chapter05.entity.TblQx;

public class RelationTest2 extends BaseHibernateDAO {
    // 在 One 一方添加数据
    public void testAddOne() {
        TblQx qx = new TblQx();
        qx.setQx(" 山南区 ");
        super.add(qx);
    }

    public static void main(String[] args) {
        new RelationTest2().testAddOne();
    }
}
```

执行结果如图 5-7 所示。

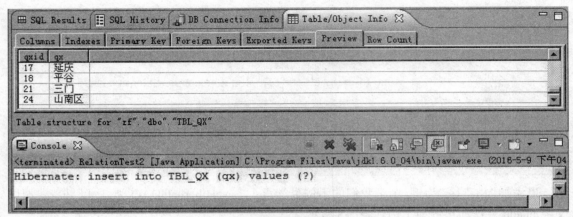

图 5-7 添加山南区

- 在 many 一方添加数据

现在我们要在"山南区"下增加 3 个街道。根据数据表的设计,在 TBL_JD 表中插入 3 条记录。使用 Hibernate,我们将采用面向对象的方式处理实体对象。首先加载得到"山南区"的实体对象,新建 jd 对象,并将"山南区"对象设置到 qx 属性中。同样,我们需要规定街道数据的主键生成策略,即配置街道实体的 id 属性,街道实体的映射文件修改如下:

```xml
<?xml version="1.0"?>
<!DOCTYPE hibernate-mapping PUBLIC "-//Hibernate/Hibernate Mapping DTD 3.0//EN"
 "http://hibernate.sourceforge.net/hibernate-mapping-3.0.dtd">

<hibernate-mapping>
    <class name="com.xtgj.j2ee.chapter05.entity.TblJd" table="TBL_JD" schema="dbo" catalog="zf" >
        <id name="jdid" column="jdid" type="int">
            <generator class="identity"></generator>
        </id>
        <property name="jd" column="jd" type="java.lang.String" not-null="false"/>
        <many-to-one name="qx" class=" com.xtgj.j2ee.chapter05.entity.TblJd" column="qxid" lazy="false"></many-to-one>
    </class>
</hibernate-mapping>
```

添加街道的源代码如示例代码 5-9 所示。

第 5 章　Hibernate 的关联映射

示例代码 5-9：添加"山南区"的街道

```java
// 在 Many 一方添加数据
    public void testAddMany(){
        TblQx qx = (TblQx)super.get(TblQx.class, 24);
        TblJd jd1 = new TblJd();
            jd1.setJd(" 和平路 ");
            jd1.setQx( qx);
            super.add(jd1);

            TblJd jd2 = new TblJd();
            jd2.setJd(" 八一路 ");
            jd2.setQx(qx);
            super.add(jd2);

            TblJd jd3 = new TblJd();
            jd3.setJd(" 五四大道 ");
            jd3.setQx(qx);
            super.add(jd3);
    }
```

执行结果如图 5-8 所示。

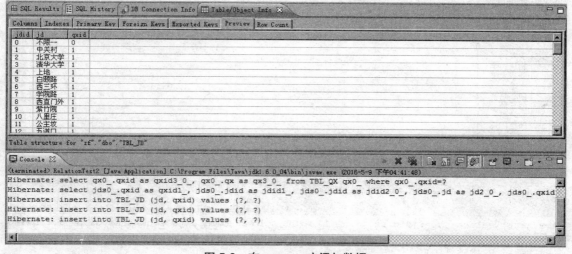

图 5-8　在 many 一方添加数据

通过执行时显示的 SQL 语句可以看到，对应执行了 3 条 insert 语句。执行结果如图 5-9 所示，可以看到 qxid 被正确设置了。

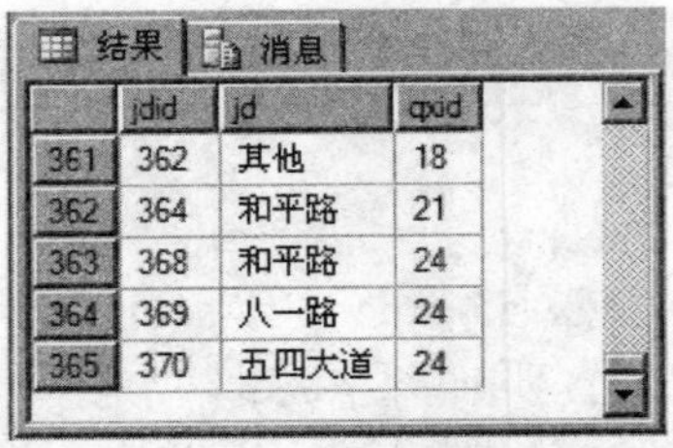

图 5-9 添加了三条记录

- 在 many 一方删除数据

现在我们要完成的任务是删除"五四大道"。根据面向对象的方式,首先加载,然后调用 session 的 delete( ) 方法。代码如示例代码 5-10 所示。

示例代码 5-10: 删除 "五四大道"
```
public void testDeleteMany(){
    TblJd jd = (TblJd)super.get(TblJd.class, 370);
    TblQx qx = (TblQx)super.get(TblQx.class, 24);
    super.delete(TblJd.class, 370);
    System.out.println(jd.getJd()+" 被删除 ");
    qx.getJds().remove(jd);
}
```

执行结果如图 5-10 所示。

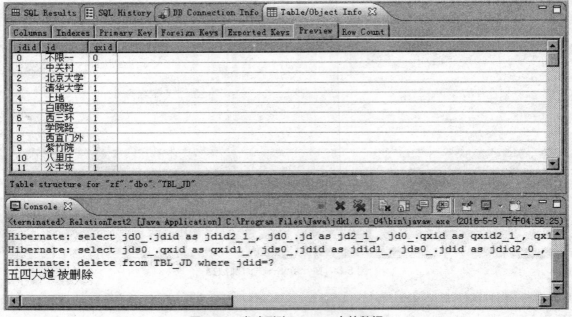

图 5-10 尝试删除 many 一方的数据

我们配置的关联是双向的,可以尝试从"八一路"对应的"山南区"那一方删除"八一路"。

如示例代码 5-11 所示。

示例代码 5-11：从"八一路"对应的"山南区"那一方进行删除
```
public void testDeleteOne() {
    TblJd jd = (TblJd) super.get(TblJd.class, 369);
    TblQx qx = (TblQx) super.get(TblQx.class, 24);
    qx.getJds().remove(jd);
    super.update(qx);
    System.out.println(jd.getJd() + " 被删除 ");
}
```

执行结果如图 5-11 所示，首先加载"山南区"对象，然后从其 jds 属性中删除"八一路"，然后更新"山南区"实体对象。

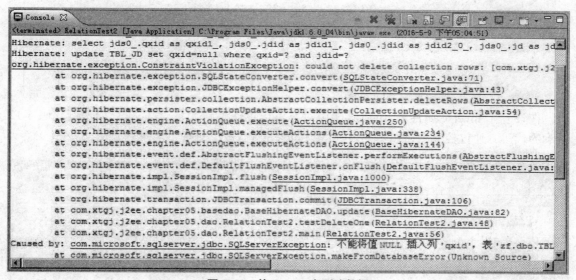

图 5-11　从 One 一方删除数据

这次执行了 update 语句，将"八一路"的 qxid 属性设成 null 了。这只是删除了关联关系，并没有删除数据库记录，这不是我们预期的效果。而且，由于违反了外键约束，操作没有成功。

原来问题出在关联关系的配置上，如示例代码 5-12 所示。配置一对多关联时，set 节点可以有一个 inverse（反转）属性，可以设为 true 或 false。为 true 时，表示由对方（one 的多方）管理双方的关联关系；为 false 时，表示自己管理双方的关联关系。一般情况下，设为 true，由多的一方管理一对多的关联关系。

示例代码 5-12：inverse（反转）属性
```
<?xml version="1.0"?>
<!DOCTYPE hibernate-mapping PUBLIC "-//Hibernate/Hibernate Mapping DTD 3.0//EN"
"http://hibernate.sourceforge.net/hibernate-mapping-3.0.dtd">
```

```xml
<hibernate-mapping >
    <class name="com.xtgj.j2ee.chapter05.entity.TblQx" table="TBL_QX" schema="dbo" catalog="zf">
        <id name="qxid" column="qxid" type="integer">
            <generator class="identity"></generator>
        </id>
        <property name="qx" column="qx" type="string" not-null="false"/>
        <set name="jds" inverse="true" lazy="false">
            <key>
                <column name="qxid" />
            </key>
            <one-to-many class="com.xtgj.j2ee.chapter05.entity.TblJd" />
        </set>
    </class>
</hibernate-mapping>
```

在设置了 inverse 属性为 true 后,即由对方(jd)管理关联关系,我们执行如示例代码 5-12 所示的代码。通过控制台可以看出,执行了 delete 语句,成功删除了记录。

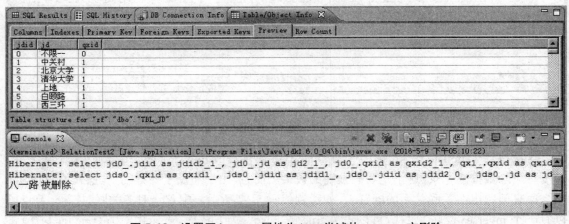

图 5-12 设置了 inverse 属性为 true 尝试从 many 一方删除

- 修改关联关系

现在我们要实现的功能是将原属于"山南区"的"和平路"划归"海淀区",即修改实体对象间的关联关系。前面我们在区县到街道的一对多关系的配置节点 <set> 上增加了 inverse 属性,并设置为 true,由多的一方负责关联关系的管理。也就是说,我们通过更新街道对象,就能更新对应的关联关系。试一下,看看是不是可以达到效果。

如示例代码 5-13 所示,首先加载相关联的 3 个实体对象:"和平路""山南区"和"海淀区"。然后从"山南区"jds 属性中删除"和平路",并添加到"海淀区"的 jds 属性中,最后,更新"和平路"对象。

示例代码 5-13:修改实体对象间的关联关系

```
public void testUpdate(){
    TblJd hepingqu = (TblJd)super.get(TblJd.class, 364);
    TblQx shannanqu = (TblQx)super.get(TblQx.class, 24);
    TblQx haidianqu = (TblQx)super.get(TblQx.class, 1);
    shannanqu.getJds().remove(hepingqu);
    haidianqu.getJds().add(hepingqu);
    hepingqu.setQx(haidianqu);
    super.update(hepingqu);
}
```

执行这段代码,看到执行了一条 update 语句,如图 5-13 所示。

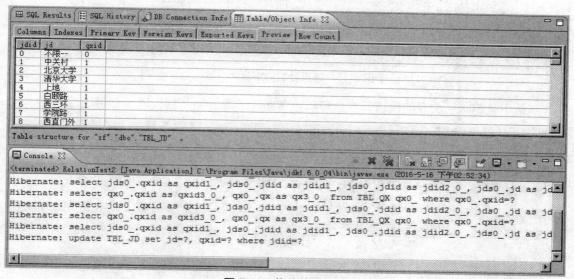

图 5-13 修改关联关系

查看数据库"和平路"对应的 qxid 字段更新了,如图 5-14 所示。

359	360	黄松峪	18
360	361	熊儿寨	18
361	362	其他	18
362	364	和平路	1

图 5-14 更改了和平路 qxid 字段

● one 一方删除数据

删除区县数据是不是直接调用 session.delete(qx) 就可以了呢?尝试运行示例代码 5-14。

示例代码 5-14：删除区县数据

```
public void testDeleteOne() {
        super.delete(TblQx.class, 21);
}
```

发现直接执行了删除 TBL_QX 记录的 delete 语句，而其下还对应有街道的数据，所以违反外键约束，抛出异常，如图 5-15 所示。

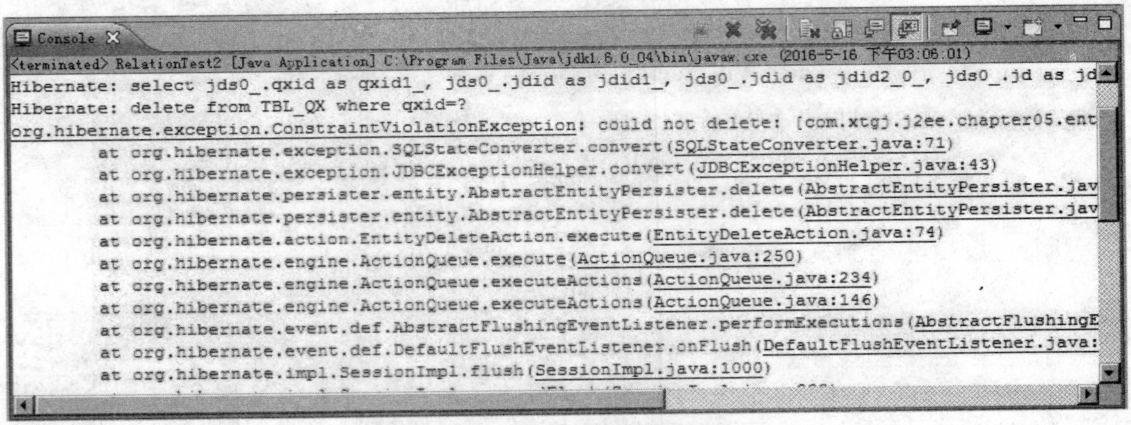

图 5-15　违反外键约束删除失败

那是不是删除了关联的街道数据就可以正常删除了呢？如示例代码 5-15 所示。

示例代码 5-15：删除区县数据

```
public void testDeleteOne() {
        TblQx qx = (TblQx) super.get(TblQx.class, 24);
        Iterator it = qx.getJds().iterator();
        while (it.hasNext()) {
                TblJd jd = (TblJd) it.next();
                super.delete(TblJd.class, jd.getJdid());
        }
        super.delete(TblQx.class, 24);
}
```

先删除了对应的街道数据，再删除区县数据，执行了多条 delete 语句，正常删除，如图 5-16 所示。

第 5 章　Hibernate 的关联映射

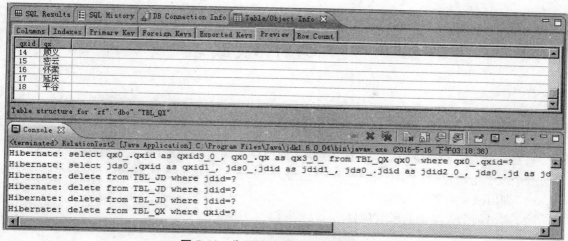

图 5-16　先删除关联的数据后删除成功

如果下属有很多条记录时，要逐条删除下属街道就太繁琐了，可以级联删除吗？Hibernate 是支持级联更新的，首先要在关联配置中设置 cascade 属性。如示例代码 5-16 所示。

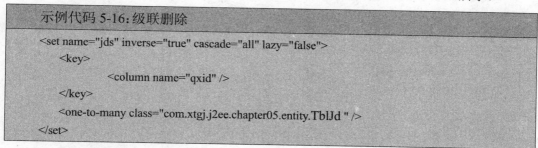

在 set 节点中增加 cascade 属性，值设为"all"。可选的值包括：
- all：对所有的操作都进行级联；
- none：对所有的操作都不进行级联；
- save-update：执行更新操作时级联；
- delete：执行删除操作时级联。

现在我们只要删除区县记录就可以自动删除关联的街道记录了。如图 5-17 所示，自动执行了从 tbl_jd 表删除记录的 delete 语句。

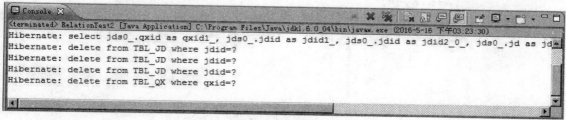

图 5-17　级联删除

5.2 多对多关联

5.2.1 配置多对多关联

某 OA 系统中要解决如下业务问题：业务部门采取项目式管理（如图 5-18 所示），每个人参与一个或多个项目；每个项目由一个或多个人完成，系统需要如下功能：

（1）列出项目的参加人员。
（2）列出某个人参加的所有项目。

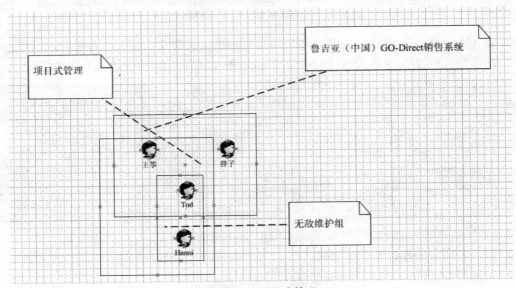

图 5-18　项目式管理

多雇员和项目使我们需要创建两个表：employee 和 project。雇员和项目间是典型的多对多关系。在设计数据库时，我们需要创建一个关联表 (r_emp_proj)，如图 5-19 所示。

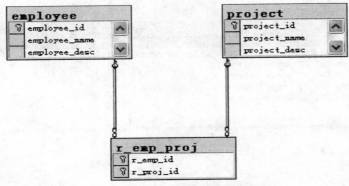

图 5-19　多对多关联的数据库结构

配置多对多节点与一对多节点类似，如示例代码 5-17 所示。首先，set 节点要指定一个 table 属性，值为关联表的表名。其下 <key> 子节点的 column 属性为关联表中关联到自己的字段名，然后是 <many-to-many> 子节点。与 <one-to-many> 类似，需要设置两个属性，class 用来设置关联属性的类型，column 属性用来设定用哪个字段作为外键去关联。最后，还要根据业务需要，将某一方的 inverse 属性设为 true。我们希望由项目 (Project) 一方管理关联，所以在 Employee 一方将 <set> 节点的 inverse 属性设为 true。

示例代码 5-17：配置多对多关联

```java
// 实体类 Employe
public class Employee implements java.io.Serializable {
    private Integer employeeId;
    private String employeeName;
    private String employeeDesc;
    private Set projects = new HashSet(0);

    public Employee() {
    }

    public Employee(String employeeName) {
        this.employeeName = employeeName;
    }

    public Employee(String employeeName, String employeeDesc, Set projects) {
        this.employeeName = employeeName;
        this.employeeDesc = employeeDesc;
        this.projects = projects;
    }

    // 省略 getter/setter
}

// 实体类 Project
public class Project implements java.io.Serializable {
    private Integer projectId;
    private String projectName;
    private String projectDesc;
    private Set members = new HashSet(0);

    public Project() {
```

```
    }
    public Project(String projectName) {
        this.projectName = projectName;
    }

    public Project(String projectName, String projectDesc, Set members) {
        this.projectName = projectName;
        this.projectDesc = projectDesc;
        this.members = members;
    }

// 省略 getter/setter

}

<!--Project-方的配置 -->
<hibernate-mapping>
    <class name="com.xtgj.j2ee.chapter05.entity.Project" table="project" schema="dbo" catalog="zf">
        <id name="projectId" type="java.lang.Integer">
            <column name="project_id" />
            <generator class="native" />
        </id>
        <property name="projectName" type="java.lang.String">
            <column name="project_name" length="50" not-null="true" />
        </property>
        <property name="projectDesc" type="java.lang.String">
            <column name="project_desc" length="500" />
        </property>
        <set name="members" table="r_emp_proj">
            <key column="r_proj_id" />
            <many-to-many class="com.xtgj.j2ee.chapter05.entity.Employee" column="r_emp_id" />
        </set>
    </class>
</hibernate-mapping>
<!--Employee-方的配置 -->
<hibernate-mapping>
```

```xml
<class name="com.xtgj.j2ee.chapter05.entity.Employee" table="employee" schema="dbo" catalog="zf">
    <id name="employeeId" type="java.lang.Integer">
        <column name="employee_id" />
        <generator class="native" />
    </id>
    <property name="employeeName" type="java.lang.String">
        <column name="employee_name" length="50" not-null="true" />
    </property>
    <property name="employeeDesc" type="java.lang.String">
        <column name="employee_desc" length="300" />
    </property>
    <set name="projects" table="r_emp_proj" inverse="true">
        <key column="r_emp_id" />
        <many-to-many class="com.xtgj.j2ee.chapter05.entity.Project" column="r_proj_id" />
    </set>
</class>
</hibernate-mapping>
```

这就完成了配置的工作,现在我们编写代码,证明配置是有效的。如示例代码 5-18 所示。

示例代码 5-18:验证查询

```java
public void testQuery() {
    Project project = (Project) super.get(Project.class, 1);
    Iterator it = project.getMembers().iterator();
    System.out.println(" 项目:" + project.getProjectName() + "\n 成员:");
    while (it.hasNext()) {
        Employee emp = (Employee) it.next();
        System.out.print(emp.getEmployeeName() + "\t");
    }
    System.out.println();
    Employee tod = (Employee) super.get(Employee.class, 2);
    Iterator it2 = tod.getProjects().iterator();
    System.out.println(" 成员:" + tod.getEmployeeName() + "\n 参与项目:");
    while (it2.hasNext()) {
        Project proj = (Project) it2.next();
        System.out.print(proj.getProjectName() + "\t");
```

```
            }
        System.out.println();
}
```

我们仅加载 Project 对象或 Employee 对象,看看是否都加载了关联的数据,如图 5-20 所示。

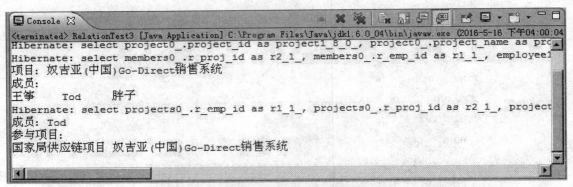

图 5-20 多对多关联示例

关联的数据都被自动加载了。使用 Hibernate,很多工作在配置文件里完成,就是这么简单。

5.2.2 通过程序维护多对多关联

现在再来编码解决下列问题:
(1)新招聘了员工:三叔,加入到国家局供应链项目。
(2)新立项项目:中教集团网联项目,三叔和胖子参加。
(3)无敌维护组项目结束,Tod 和 Hanni 退出项目组。

前面已经做好了配置,编码工作就变得很轻松了。首先看第一个问题:新添雇员"三叔",加入到"国家局供应链项目"。首先,创建新 Employee 对象,加载国家局项目对象,然后设置相互的关联,最后执行持久化操作。先对新建对象调用 session.save() 方法,然后对需要更新的对象调用 session.update() 方法。(我们使用 BaseHibernateDAO 对这些操作进行了封装,直接调用 BaseHibernateDAO 的相应方法即可,还免去了繁琐的 try...catch 和事务控制语句)。如示例代码 5-19 所示,执行了两条 insert 语句,一条插入了 Employee 记录,一条在关系表插入了记录,建立了"三叔"到"国家局供应链项目"间的关联关系。

示例代码 5-19:新添雇员测试

```
public void testAddEmployee(){
    Employee sanshu = new Employee();
    sanshu.setEmployeeName(" 三叔 ");
    Project guojiaju = (Project)this.get(Project.class, 2);
    sanshu.getProjects().add(guojiaju);
```

```
            guojiaju.getMembers().add(sanshu);
            super.add(sanshu);
            super.update(guojiaju);
    }
```

执行结果如图 5-21 所示。

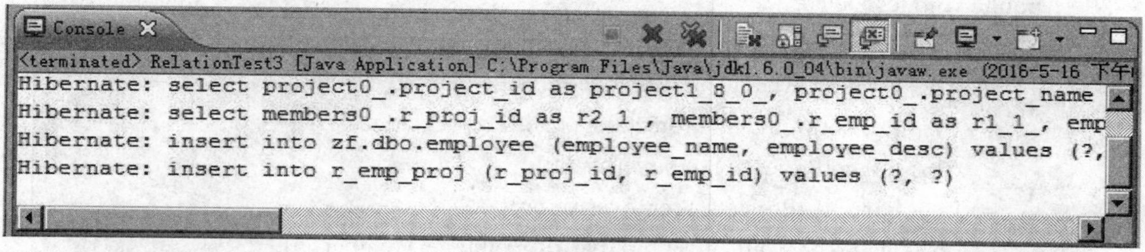

```
Hibernate: select project0_.project_id as project1_8_0_, project0_.project_name
Hibernate: select members0_.r_proj_id as r2_1_, members0_.r_emp_id as r1_1_, emp
Hibernate: insert into zf.dbo.employee (employee_name, employee_desc) values (?,
Hibernate: insert into r_emp_proj (r_proj_id, r_emp_id) values (?, ?)
```

图 5-21 新添雇员

第二个问题：新建立项目，三叔和胖子参加。这个问题同上一个问题类似，步骤也类似。加载或新建实体对象，然后编码维护双方关联关系，最后调用持久化方法。代码如示例代码 5-20 所示。

示例代码 5-20：新建立项目测试
```
public void testAddProject(){
    Employee pangzi = (Employee)this.get(Employee.class, 3);
    Employee sanshu = (Employee)this.get(Employee.class, 5);
    Project wangyin = new Project();
    wangyin.setProjectName(" 中教集团网银项目 ");
    wangyin.getMembers().add(pangzi);
    wangyin.getMembers().add(sanshu);
    super.update(pangzi);
    super.update(sanshu);
    super.add(wangyin);
}
```

执行结果如图 5-22 所示。

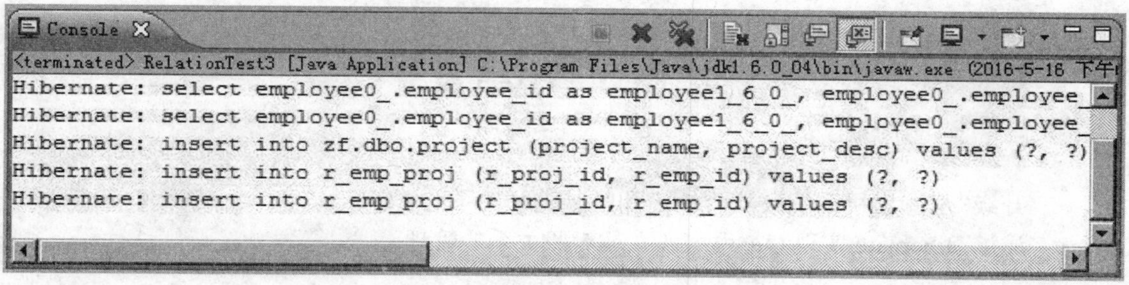

```
Hibernate: select employee0_.employee_id as employee1_6_0_, employee0_.employee_
Hibernate: select employee0_.employee_id as employee1_6_0_, employee0_.employee_
Hibernate: insert into zf.dbo.project (project_name, project_desc) values (?, ?)
Hibernate: insert into r_emp_proj (r_proj_id, r_emp_id) values (?, ?)
Hibernate: insert into r_emp_proj (r_proj_id, r_emp_id) values (?, ?)
```

图 5-22 新建立项目

第三个问题:无敌维护组项目结束,Tod 和 Hanni 退出项目组。首先加载相关对象,然后通过编码维护关联关系,删除关联时使用集合对象的 remove() 方法。可以采用"weihu.getMembers().removeAll(weihu.getMembers());"的编程技巧将对象的集合属性清空,最后执行持久化操作,删除操作放在最后,以便删除时关联关系已经被清除。如示例代码 5-21 所示。

示例代码 5-21:删除项目测试

```java
public void testDelete(){
    Employee tod = (Employee)this.get(Employee.class, 2);
    Employee hanni = (Employee)this.get(Employee.class, 4);
    Project weihu = (Project)super.get(Project.class, 4);
    tod.getProjects().remove(weihu);
    hanni.getProjects().remove(weihu);
    weihu.getMembers().removeAll(weihu.getMembers());
    super.update(tod);
    super.update(hanni);
    super.delete(Project.class, 3);
}
```

执行结果如图 5-23 所示。

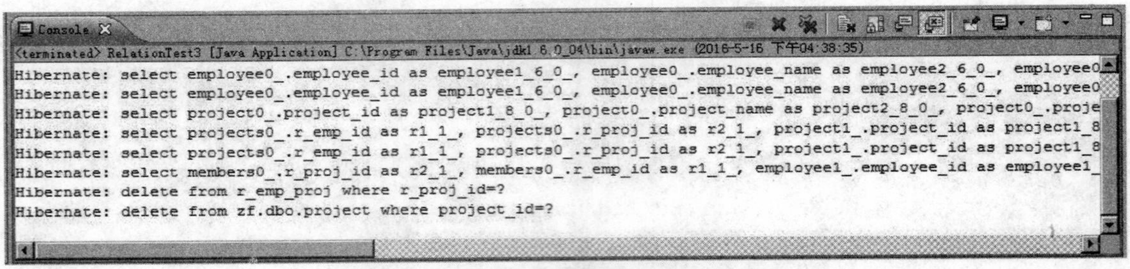

图 5-23 删除多对多的关联关系

使用 Hibernate,我们采用面向对象的方式操作数据,管理对象间的关联关系时也是采用这样的思路。首先要做的是做好关联的配置,通过配置关联,Hibernate 就会自动帮我们加载关联的数据,以及进行其他持久化操作。在编写代码的时候,我们心里只有对象,再也不用去管数据库、表和 SQL 语句了。

5.3 小结

- ✓ 升级"房屋出租系统":实现街道和区县关联关系的管理。
- ✓ 通过 Hibernate 实现 OA 系统和人员对照关系的管理。

5.4 英语角

employee　　　雇员
update　　　　更新

5.5 作业

实现一个网上书店中图书分类级联菜单的设计。菜单项应该来自于数据库，数据库中包含图书主分类表和次分类表，当页面加载时显示图书主分类菜单项列表，当用户点击某一个主菜单项时，在该项菜单的下方展开并显示相应的次分类菜单项列表，当用户再次点击主菜单项时则收缩子菜单项列表。

5.6 思考题

Hibernate 关联映射主要有哪几种？

5.7 学员回顾内容

1. 单向一对多关联。
2. 单向多对一关联。
3. 双向多对一关联。

第 6 章 Hibernate 查询

学习目标

- ✧ 掌握 HQL 查询。
- ✧ 掌握 Criteria 查询。

课前准备

在互联网上查阅 HQL 查询的相关知识

本章简介

前面我们学习了 Hibernate 基础知识，以及如何使用 Hibernate 管理对象间的关联关系。在第 4 章，我们只介绍了如何使用 Hibernate 完成新建、删除、更新以及加载对象数据的方法，没有介绍如何使用 Hibernate 进行查询操作，本次课我们将完成这部分知识的学习。

Hibernate 支持两种主要的查询方式：HQL（Hibernate Query Languege，Hibernate 查询语言）查询和 Criteria 查询。HQL 是一种面向对象的查询语言，其中没有表和字段的概念，只有类、对象和属性的概念，这点需要大家好好体会。Criteria 查询又称为"对象查询"，它用面向对象的方式将构造查询的过程做了封装。其中 HQL 是应用较为广泛的方式，也是我们介绍的重点。

6.1 Hibernate 查询语言

6.1.1 为什么使用 HQL

回想一下使用 JDBC 时，我们的查询方法是什么样子的。图 6-1 中展示了一个典型的使用 JDBC 的查询方法，首先是繁复的 SQL 语句，然后还要把查询结果以对象的形式封装，放到集合对象中。这种方法代码繁琐，容易出错。

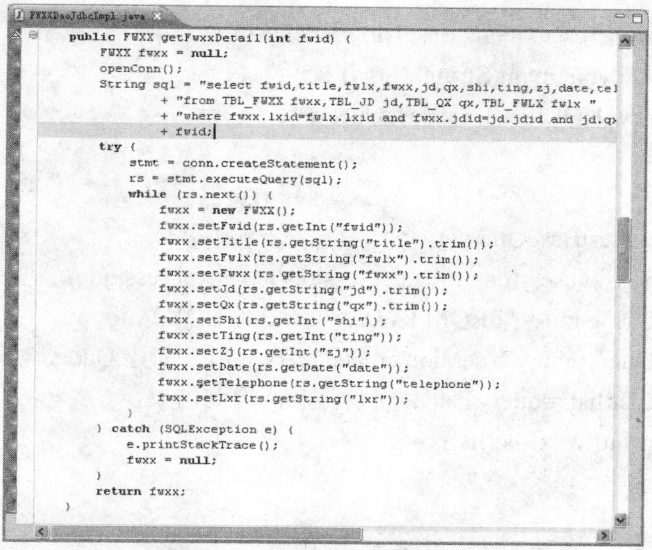

```java
public FWXX getFwxxDetail(int fwid) {
    FWXX fwxx = null;
    openConn();
    String sql = "select fwid,title,fwlx,fwxx,jd,qx,shi,ting,zj,date,tel
            + "from TBL_FWXX fwxx,TBL_JD jd,TBL_QX qx,TBL_FWLX fwlx "
            + "where fwxx.lxid=fwlx.lxid and fwxx.jdid=jd.jdid and jd.q
            + fwid;
    try {
        stmt = conn.createStatement();
        rs = stmt.executeQuery(sql);
        while (rs.next()) {
            fwxx = new FWXX();
            fwxx.setFwid(rs.getInt("fwid"));
            fwxx.setTitle(rs.getString("title").trim());
            fwxx.setFwlx(rs.getString("fwlx").trim());
            fwxx.setFwxx(rs.getString("fwxx").trim());
            fwxx.setJd(rs.getString("jd").trim());
            fwxx.setQx(rs.getString("qx").trim());
            fwxx.setShi(rs.getInt("shi"));
            fwxx.setTing(rs.getInt("ting"));
            fwxx.setZj(rs.getInt("zj"));
            fwxx.setDate(rs.getDate("date"));
            fwxx.getTelephone(rs.getString("telephone"));
            fwxx.setLxr(rs.getString("lxr"));
        }
    } catch (SQLException e) {
        e.printStackTrace();
        fwxx = null;
    }
    return fwxx;
}
```

图 6-1 使用 JDBC 的查询方法

HQL 是 Hibernate 提供的一种面向对象的查询语言，使用 HQL 可以避免使用 JDBC 查询的一些弊端。首先，不用再编写繁复的 SQL 语句，我们将针对实体类及其属性进行查询。其次，查询结果是直接存放在 List 中的对象，不需要再次封装。而且 HQL 还独立于数据库，对不同的数据库根据 Hibernate dialect 属性的配置自动生成不同的 SQL 语句执行。

下面我们就看看 HQL 是怎样使用的。

6.1.2 如何使用 HQL

使用 HQL 需要四步：

（1）得到 session。

（2）编写 HQL 语句。

（3）创建 Query 对象。

（4）执行查询，得到结果。

如示例代码 6-1 所示。

示例代码 6-1：使用 HQL

```java
package com.xtgj.j2ee.chapter06.hql;

import java.util.Iterator;
import java.util.List;
import org.hibernate.Query;
import org.hibernate.Session;

import com.xtgj.j2ee.chapter06.basedao.BaseHibernateDAO;
import com.xtgj.j2ee.chapter06.entity.Fwxx;
```

```java
public class HQLTest extends BaseHibernateDAO {
    public static void main(String[] args) {
            new HQLTest().testBaseQuery();
    }

    public void testBaseQuery() {
            Session session = this.getSession(); // 得到 session
            String hql = "FROM FWXX"; // 编写 HQL 语句
            Query query = session.createQuery(hql);// 创建 Query 对象
            List list = query.list(); // 执行查询,得到结果
            printFwxxList(list);
    }

    private void printFwxxList(List list) {
            Iterator it = list.iterator();
            while (it.hasNext()) {
                    Fwxx item = (Fwxx) it.next();
                    System.out.println(item.getLxr() + "\t[" + item.getDate() + "]\t"
                                    + item.getTitle() + "(" + item.getZj() + ")");
            }
    }
}
```

这四个步骤是使用 HQL 查询的"套路代码",要记下来达到可以默写的程度哦!我们重点关注 HQL 的语法。在代码 6-1 中,HQL 语句为"FROM FWXX"。其中 FWXX 并非表名,而是类名。既然是类名,就要区分大小写,因此写成"FROM fwxx"是不可以的。但 FROM 关键字不区分大小写,所以也可以写成"from com.xtgj.j2ee.chapter06.entity.FWXX"。类名可以只是类名,也可以包含包名,而且与 SQL 语句不同,可以没有 SELECT 子句。

在 HQL 中,可以给类名指定别名,上面的 HQL 语句可以写成:

```java
String hql = "SELECT fw FROM FWXX AS fw";
```

SELECT 子句中内容是 fw,表示查询所有的房屋信息数据,以 FWXX 对象列表的方式返回。其中 fw 是 FWXX 的别名,通过 AS 关键字指定,AS 关键字可以省略。这条 HQL 语句执行结果如图 6-2 所示。

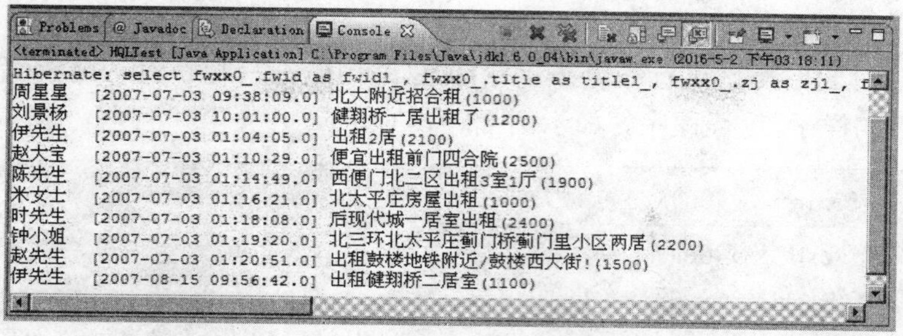

第 6 章　Hibernate 查询

[图 6-2 查询房屋信息截图]

图 6-2　查询房屋信息

HQL 拥有丰富的语法和强大功能，下面我们逐一学习。

6.1.3　HQL 的丰富功能

● 属性查询

假如我们不想查询整个对象的数据，只需要房屋出租信息中的发布日期和标题，HQL 语句怎么写？如示例代码 6-2 所示。

```
示例代码 6-2：查询部分属性
public void testPropertyQuery(){
        Session session = this.getSession();
        String hql = "select fw.date, fw.title from Fwxx fw";
        Query query = session.createQuery(hql);
        List list = query.list();
        Iterator it = list.iterator();
        while(it.hasNext()){
                Object[] arr = (Object[])it.next();
                System.out.println(arr[0] + "\t" + arr[1]);
        }
}
```

HQL 也支持只查询对象中的某几个属性。查询结果还是保存在 List 中，不过每条数据体现为一个 Object 数组的形式。运行结果如图 6-3 所示。

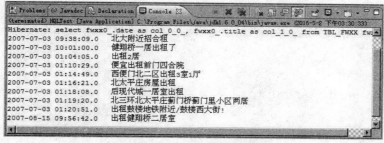

图 6-3　查询房屋信息的部分属性

● 参数查询

假如我们想对 title 属性进行模糊查询，HQL 语句怎么写呢？将示例代码 6-1 中的 HQL 语句修改成如下的样子，运行程序。如示例代码 6-3 所示。

```
示例代码 6-3：参数查询
public void testLikeQuery(){
        Session session = this.getSession();
        String hql = "from Fwxx fw where fw.title like '% 健翔桥 %'";
        Query query = session.createQuery(hql);
        List list = query.list();
        Iterator it = list.iterator();
        while (it.hasNext()) {
                Fwxx item = (Fwxx) it.next();
                System.out.println(item.getLxr() + "\t[" + item.getDate() + "]\t"
                        + item.getTitle() + "(" + item.getZj() + ")");
        }
}
```

运行结果如图 6-4 所示。可见 HQL 语句支持 where 子句，支持 like 关键字，也支持"%"通配符。

图 6-4　模糊查询

可是这样拼装 HQL 语句容易带来安全隐患，有类似 PreparedStatement 预处理的做法吗？答案是肯定的。具体做法如示例代码 6-4 所示。

```
示例代码 6-4：HQL 查询中的预处理
public void testParamQuery() {
        Session session = this.getSession();
        String hql = "from Fwxx fw where fw.title like ?";
        Query query = session.createQuery(hql);
        query.setString(0, "% 健翔桥 %");
        List list = query.list();
        printFwxxList(list);
}
```

首先使用"?"作占位符，然后通过 Query 的 setString() 方法逐个设置参数的值。Query 还提供 setLong()、setDouble() 和 setDate() 等方法用于设置不同类型的参数值。需要注意的是：

必须保证每个占位符都设置了参数值,设置参数值时,下标从 0 开始。当参数个数多的时候,使用占位符这种方式也有一些缺点,比如:代码可读性下降,需要仔细比对才知道哪个设置参数的语句对应到哪个占位符上,给查找错误带来麻烦。还有就是将参数的下标硬编码,参数顺序调整,代码也需要跟着调整,很不方便。我们可以使用"命名参数查询"解决这些问题。

现在,我们需要定义一个查询方法:public List searchByZj(int zjl,intzj2),传入租金的范围,返回租金符合这个范围的租房信息。如示例代码 6-5 所示。

示例代码 6-5:HQL 查询中的命名参数查询

```
public List searchByZj(int zj1, int zj2){
    Session session = this.getSession();
    String hql = "from Fwxx fw " +
            "where fw.zj between ? and ?";
    Query query = session.createQuery(hql);
    query.setInteger(0, zj1);
    query.setInteger(1,zj2);
    return query.list();
}
```

命名参数采用":参数名"的格式定义参数,然后通过名称逐个设置参数的值。这样程序的可读性和可维护性都增强了。

在代码 6-5 中,HQL 语句的 where 子句使用了">="和"<="运算符。HQL 支持 >、<、=、>=、<=、<>(不等于)和 is null 等运算符,支持 and、or、not 和括号,而且还支持 in 和 between。那么,使用 between 怎么改写代码 6-5 中的 HQL 语句呢?如示例代码 6-6 所示的 HQL 语句就可以了。

示例代码 6-6:HQL 查询中的命名参数 between 查询

```
public List searchByZj(int zj1, int zj2){
    Session session = this.getSession();
    String hql = "from Fwxx fw " +
            "where fw.zj between :zj1 and :zj2";
    Query query = session.createQuery(hql);
    query.setInteger("zj1", zj1);
    query.setInteger("zj2",zj2);
    return query.list();
}
```

执行示例代码 6-6,得到的结果如图 6-5 所示。

```
Hibernate: select fwxx0_.fwid as fwid1_, fwxx0_.title as title1_, fwxx0_.zj as zj1_, fwxx0_
伊先生    [2007-07-03 01:04:05.0]  出租2居 (2100)
赵大宝    [2007-07-03 01:10:29.0]  便宜出租前门四合院 (2500)
时先生    [2007-07-03 01:18:08.0]  后现代城一居室出租 (2400)
钟小姐    [2007-07-03 01:19:20.0]  北三环北太平庄蓟门桥蓟门里小区两居 (2200)
```

图 6-5　between-and 查询

- 关联查询

如果我们想查询联系人为"伊先生"的房屋信息都分布在那些街道，该怎么办呢？

上述问题查询的结果是街道的列表，我们知道，查询所有街道的 HQL 语句是"select jd from JD jd"，但怎么编写查询语句实现问题要求的 HQL 呢？

联系人是房屋信息的属性，所以我们需要关联 JD 和 Fwxx 两类数据进行查询。如示例代码 6-7 所示。

示例代码 6-7：HQL 查询中的关联查询

```
public void testAssociatedQuery() {
        Session session = this.getSession();
        String hql = " select jd " + "from JD jd,F fw "
                   + "where fw.jd=jd and fw.lxr=' 伊先生 '";
        Query query = session.createQuery(hql);
        List list = query.list();
        printJdList(list);

}

public void printJdList(List list) {
        Iterator it = list.iterator();
        while (it.hasNext()) {
                JD item = (JD) it.next();
                System.out.println(item.getJd());
        }
}
```

示例代码 6-7 中，我们关联了 JD 和 Fwxx 两类数据，关联的条件是"fw.jd=jd"，我们注意到这个关联条件等式的两端都是对象，Hibernate 通过两个对象主键值是否相等判断两个对象是否相等。观察示例代码 6-5 执行时输出的 SQL 语句可以注意到，"fw.jd=jd"关联条件最终转换成"fwxx1_.jdid=jd0_.jdid"，通过 id 去关联。我们编写 HQL 时，只要按照面向对象的思维去考虑和设置关联条件就可以了。本示例执行结果如图 6-6 所示。

图 6-6 关联查询

- 分页查询

分页是我们一直关注的问题,在 HQL 中实现分页查询方法如下:

```
public List search(int pageNo,int pageSize);
```

分页查询是系统中常常需要的功能,当数据库记录条数很多的时候,每次只加载一页的数据也是不得不采取的策略。现在我们就看看如何使用 Hibernate 来实现,如示例代码 6-8 所示。

示例代码 6-8:HQL 查询中的分页查询

```
public List search(int pageNo, int pageSize) {
    Session session = this.getSession();
    String hql = "from Fwxx fw order by fw.fwid asc";
    Query query = session.createQuery(hql);
    int firstResultIndex = pageSize * (pageNo - 1);
    query.setFirstResult(firstResultIndex);
    query.setMaxResults(pageSize);
    return query.list();
}
```

如示例代码 6-8 所示,首先通过"order by"指定返回结果的排序规则,然后通过 query.setFirstResult 和 query.setMaxResults 设置从第几条数据开始,共返回多少条数据。现在我们按照如下所示的语句调用示例代码 6-6 中的 search 方法即可得到每页 15 条数据时第 3 页的数据。

```
List ret = search(2,4);
```

执行上述示例代码,得到的结果如图 6-7 所示。

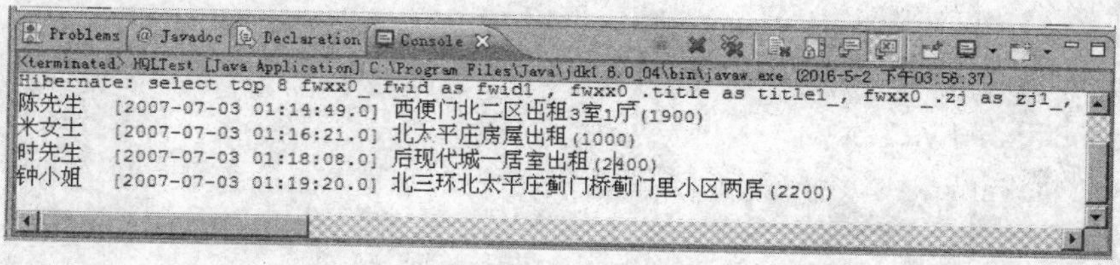

图 6-7 分页查询

- 统计函数

一般,实现分页功能时我们需要知道总的数据条数。那么,通过 HQL 查询怎么得到呢?如示例代码 6-9 所示。

示例代码 6-9：HQL 查询中的统计函数

```
public int getTotalCount(){
        Session session = this.getSession();
        String hql = "select count(fw) from FWXX fw";
        Query query = session.createQuery(hql);
        int count = Integer.parseInt(query.uniqueResult().toString());
        return count;
}
```

如示例代码 6-9 所示，可以通过 count 函数得到数据条数。HQL 支持的统计函数还有 min()、max()、avg() 等。当查询结果只有 1 条的时候，我们可以通过 query.uniqueResult() 方法直接获得结果。

执行上述示例代码，得到的结果如图 6-8 所示。

图 6-8　统计查询

现在我们已经可以使用 HQL 实现很复杂的查询了。接下来，我们利用 HQL 完成一个简单的应用。

问题：定义一个查询方法 public List search(FWXX condition)，同时支持以下 3 项功能。
（1）对 title 模糊查询。
（2）对租金使用一个范围查询。
（3）对房屋类型精确查询。

分析：为了保存查询条件，首先在 Fwxx 类内增加 3 个属性，Fwxx 类的示例代码 6-10 如下。

示例代码 6-10：Fwxx 类

```
package com.xtgj.j2ee.chapter06.entity;

public class Fwxx{
    private int fwid;
    private int zj;
    private int shi;
    private int ting;
    private String title;
    private String fwxx;
    private String date;
    private String telephone;
```

```java
        private String lxr;
        private USER user;
        private JD jd;
        private FWLX fwlx;

        private Double zj1;// 存放查询条件用
        private Double zj2;// 存放查询条件用
        private Integer[] lxIds =null;// 存放查询条件用

        public Double getZj1() {
                return zj1;
        }

        public void setZj1(Double zj1) {
                this.zj1 = zj1;
        }

        public Double getZj2() {
                return zj2;
        }

        public void setZj2(Double zj2) {
                this.zj2 = zj2;
        }

        public Integer[] getLxIds() {
                return lxIds;
        }

        public void setLxIds(Integer[] lxIds) {
                this.lxIds = lxIds;
        }

        // 省略其他属性 getter/setter
}
```

代码如示例代码 6-11 所示。其中 arrayToString 方法用于将数组数据转换为逗号分隔的字符串，请自行实验。

示例代码 6-11：复杂查询

```java
public List search(Fwxx condition) {
    Session session = this.getSession();
    String hql = "from FWXX fw where 1=1 ";
    if (null != condition) {
        if (condition.getTitle() != null
                && !condition.getTitle().equals("")) {
            hql += " and fw.title like '%" + condition.getTitle() + "%'";
        }
        if (condition.getZj1() != null && condition.getZj1() != 0) {
            hql += " and fw.zj >= " + condition.getZj1();

        }
        if (condition.getZj2() != null && condition.getZj2() != 0) {
            hql += " and fw.zj <= " + condition.getZj2();

        }
        if (condition.getLxIds() != null && condition.getLxIds().length > 0) {
            hql += " and fw.fwlx.lxid in("
                    + arrayToString(condition.getLxIds())
                    + ")";
            ;
        }
    }
    Query query = session.createQuery(hql);
    return query.list();
}

private String arrayToString(Integer[] lxIds) {
    String s = "";
    for (int i = 0; i < lxIds.length - 1; i++) {
        s += lxIds[i] + ",";
    }
    s += lxIds[lxIds.length - 1];
    return s;
}
```

在项目中,我们也常常需要定义类似的查询方法。此时的代码已经比采用 JDBC 的代码精简了很多。注意查询条件中的 fw.fwlx.1xid,只有面向对象查询语言中才可以这样做。

编写测试代码如示例代码 6-12 所示。

示例代码 6-12:测试代码

```
public static void main(String[] args) {
    Fwxx fw = new Fwxx();
    fw.setZj1(new Double(1000));
    fw.setZj2(new Double(3000));
    fw.setTitle(" 北 ");
    fw.setLxIds(new Integer[]{3,4});
    new HQLTest().printFwxxList(new HQLTest().search(fw));
}
```

执行上述代码,得到如图 6-9 所示的结果。

图 6-9 多条件复杂查询

6.2 Criteria 查询

6.2.1 为什么使用 Criteria 查询

示例代码 6-11 虽然比采用 JDBC 的代码已经有了很大进步,但代码还是很啰嗦,而且不方便使用参数查询,安全性和运行效率都会受影响。有什么办法可以改进呢?

Criteria 查询(又称作对象查询)采用面向对象的方式(主要是 Criteria 对象)封装查询条件,并提供了 Restrictions 等类辅助查询,可以使编写查询代码更方便,而且代码更易读。现在我们就来看看具体是怎么做的。

6.2.2 如何使用 Criteria 查询

如示例代码 6-13 所示,使用 Criteria 首先创建 Criteria 对象,与创建 Query 对象的语法很相似,但需要传入的参数是对应实体类的类型对象。然后使用 Restrictions 对象的静态方法(包括 like、ge、1e、in 等)构造查询条件,并添加进 Criteria 对象中。其中,like 方法构造一个模糊查询的查询条件,第一个参数是属性名,第二个参数是查询条件的值,第三个参数是关键字匹配方式,可选的有 MatchMode.ANYWHERE、MatchMode.END、MatchMode.START 和

MatchMode.EXACT(精确查询)。ge 表示构造大于等于的查询条件，le 表示构造小于等于的查询条件。最后，还可以通过 Criteria 的 addOrder() 方法指定查询的排序方式。同样，分页查询的两个方法 Criteria 对象也支持，所以一般构造复杂的查询方法时，我们采用 Criteria 方式。

示例代码 6-13：Criteria 查询

```java
package com.xtgj.j2ee.chapter06.hql;

import java.util.Iterator;
import java.util.List;
import org.hibernate.Criteria;
import org.hibernate.Session;
import org.hibernate.criterion.MatchMode;
import org.hibernate.criterion.Order;
import org.hibernate.criterion.Restrictions;
import com.xtgj.j2ee.chapter06.entity.FWXX;

public class CriteriaTest extends BaseHibernateDAO {

    public static void main(String[] args) {
        Fwxx fw = new Fwxx ();
        fw.setZj1(new Double(2000));
        fw.setZj2(new Double(3000));
        new CriteriaTest().printFwxxList(new CriteriaTest().search(fw));
    }

    public List search(Fwxx condition) {
        Session session = this.getSession();
        Criteria c = session.createCriteria(Fwxx.class);
        if (null != condition) {
            if (condition.getTitle() != null
                    && !condition.getTitle().equals("")) {
                c.add(Restrictions.like("title", condition.getTitle(),
                        MatchMode.ANYWHERE));
            }
            if (condition.getZj1() != null && condition.getZj1() != 0) {
                c.add(Restrictions.ge("zj", condition.getZj1().intValue()));
            }
            if (condition.getZj2() != null && condition.getZj2() != 0) {
                c.add(Restrictions.le("zj", condition.getZj2().intValue()));
```

```
                    }
                    if (condition.getLxIds() != null && condition.getLxIds().length > 0) {
                        c.add(Restrictions.in("lx.lxid", condition.getLxIds()));
                    }
                }
                c.addOrder(Order.asc("fwid"));
                return c.list();
            }

            private void printFwxxList(List list) {
                Iterator it = list.iterator();
                while (it.hasNext()) {
                    FWXX item = (FWXX) it.next();
                    System.out.println(item.getLxr() + "\t[" + item.getDate() + "]\t"
                            + item.getTitle() + "(" + item.getZj() + ")");
                }
            }
        }
```

执行上述代码,得到如图 6-10 所示的结果。

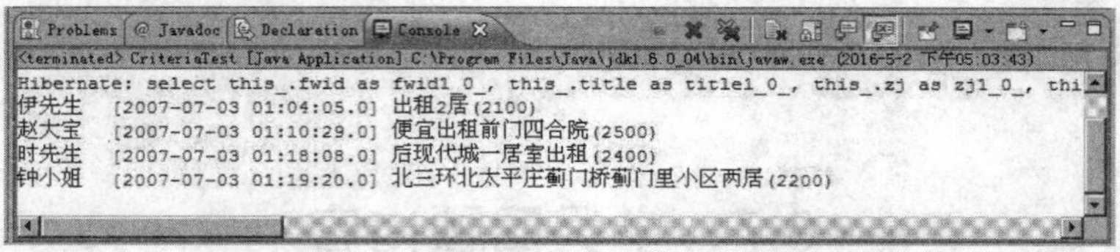

图 6-10 条件查询

至此,我们学习了 Hibenate 的基础知识、关联和查询,但真正做到使用熟练还需要多编写代码,使用这些知识解决各种各样的项目需求。只有在实践中不断犯错误、积累经验,才能真正地学会一门技术。

6.3 小结

✓ Hibernate 支持两种主要的查询方式:HQL 查询和 Criteria 查询。
✓ HQL 是一种面向对象的查询语言,支持属性查询、参数查询、关联查询、分页查询等特性,还支持统计函数。
✓ 执行 HQL 语句首先要创建 Query 对象,Query 对象封装了查询所需要的大部分操作

（设置参数，执行查询，设置返回数据的开始位置和总的返回条数等）。

✓ 复杂查询使用 Criteria 将更方便。Criteria 采用面向对象的方式对查询条件的组装过程进行了封装。Restrictions 提供了很多静态方法用于构造查询条件。

6.4 英语角

query　　　　　　查询
criteria　　　　　　条件

6.5 作业

1. 使用 HQL 查询，完成房屋租赁系统中房屋信息的高级查询功能。
2. 使用 Criteria 查询，完成房屋租赁系统中房屋信息的高级查询功能。

6.6 思考题

哪种情况使用 HQL 查询？哪种情况使用 Criteria 查询？

6.7 学员回顾内容

1. HQL 查询。
2. Criteria 查询。

第 7 章　Hibernate 高级特性

学习目标

- ◇ 了解 Hibernate 检索策略。
- ◇ 掌握 Hibernate 延迟加载。
- ◇ 了解 Hibernate 缓存机制。
- ◇ 掌握 Hibernate 缓存配置方式。

课前准备

在互联网上了解 Hibernate 的缓存机制和配置。

本章简介

在前面的内容中，我们探讨了 Hibernate 的基础使用技术。通过对这些基础技术的掌握，我们即可开始进行基于 Hibernate 的持久层开发。然而，在这些应用技术之后，存在着怎样的运行机制以及其内部实现方式对应用层可能产生怎样的影响，则是下面需要关注的问题。

7.1　Hibernate 检索策略

7.1.1　类级别检索策略和关联级别检索策略

Hibernate 的检索策略包括类级别检索策略和关联级别检索策略。

类级别检索策略又分为立即检索和延迟检索两类，默认的检索策略是立即检索。在 Hibernate 映射文件中，通过在 <class> 上配置 lazy 属性来确定检索策略。对于 Session 的检索方式，类级别检索策略仅适用于 load 方法，也就说，对于 get 检索和 query 检索，持久化对象都会被立即加载而不管 lazy 的值是 false 还是 true。一般来说，我们检索对象就是要访问它，因此立即检索是通常的选择。由于 load 方法在检索不到对象时会抛出异常（立即检索的情况下），因此编者不建议使用 load 检索。而且，由于 <class> 中的 lazy 属性还影响到多对一及一对一的检索策略，因此使用 load 方法就更没必要了。

关联级别检索策略又分为立即检索、延迟检索和迫切左外连接检索三类。对于关联级别检索，亦可按照一对多和多对多、多对一和一对一两种情况讨论：

一对多和多对多关联关系一般使用 <set> 节点配置。<set> 节点中有 lazy 和 outer-join 属性，它们的不同取值决定了不同的检索策略。

（1）立即检索：这是一对多默认的检索策略，此时 <set> 节点的"lazy=false"，"outer-join=false"。尽管这是默认的检索策略，但如果关联的集合是无用的，那么就不要使用这种检索方式。

（2）延迟检索：此时"lazy=true"，"outer-join=false"（outer-join=true 是无意义的），这是优先考虑的检索方式。

（3）迫切左外连接检索：此时"lazy=false"，"outer-join=true"，这种检索策略只适用于依靠 id 检索方式（例如 load、get），而不适用于 query 的集合检索（query 检索会采用立即检索策略）。相比于立即检索，这种检索策略减少了一条 sql 语句，但在 Hibernate 中，只能有一个 <set> 节点配置成"outer-join=true"。

多对一和一对一检索策略一般使用 <many-to-one>、<one-to-one> 节点配置。<many-to-one> 中需要配置的属性是 outer-join，同时还需要配置 one 端关联的 <class> 节点的 lazy 属性（注意：配置的不是 <many-to-one> 中的 lazy 哦），它们的组合后的检索策略如下：

（1）<many-to-one> 中的"outer-join=auto"：这是默认值，如果 one 端关联的 <class> 的"lazy=true"则为延迟检索，如果 one 端关联的 <class> 的"lazy=false"则为迫切左外连接检索。

（2）<many-to-one> 中的"outer-join=true"：无关于 one 端关联的 <class> 的 lazy 属性，都为迫切左外连接检索。

（3）<many-to-one> 中的"outer-join=false"：如果 one 端关联的 <class> 的"lazy=true"则为延迟检索，否则即为立即检索。

可以看到，在默认的情况下（"outer-join=auto"，"lazy=false"），对关联的 one 端对象 Hibernate 采用迫切左外连接检索。很多情况下，并不需要加载 one 端关联的对象（大多数情况需要的仅仅是关联对象的 id），另外，如果关联对象也采用了迫切左外连接检索，就会出现 select 语句中有多个外连接表，如果个数过多的话就会影响检索性能，这也是为什么 Hibernate 会通过 hibernate.max_fetch_depth 属性来控制外连接的深度。对于迫切左外连接检索，query 的集合检索并不适用，它会采用立即检索策略。

对于检索策略，需要根据实际情况进行选择。对于立即检索和延迟检索，它们的优点在于 select 语句简单（每张表一条语句）、查询速度快；缺点在于关联表时需要多条 select 语句，增加了访问数据库的频率。因此，在选择即检索和延迟检索时，可以考虑使用批量检索策略来减少 select 语句的数量（配置 batch-size 属性）。对于迫切左外连接检索，优点在于 select 较少，但缺点是 select 语句的复杂度提高，多表之间的关联会是很耗时的操作。可以根据需要在程序里显示指定的检索策略（可能经常需要在程序中显示指定迫切左外连接检索）。为了清楚检索策略的配置效果如何，可以配置 show_sql 属性查看程序运行时 Hibernate 执行的 sql 语句。

7.1.2　Hibernate 延迟加载机制

在上一小节里我们简单地介绍了 Hibernate 的检索策略，在这一节我们将详细介绍 Hibernate 延迟加载的问题。延迟加载机制是为了避免一些无用的性能开销而提出来的，所谓延迟加载就是在真正需要数据的时候，才执行数据加载操作。在 Hibernate 中提供了对实体对象的延迟加载以及对集合的延迟加载,另外在 Hibernate 3.0 中还提供了对属性的延迟加载。下面我们就分别介绍这些种类的延迟加载的细节。

● 实体对象的延迟加载

如果想对实体对象使用延迟加载,必须要在实体的映射配置文件中进行相应的配置,如示例代码 7-1 所示。

> 示例代码 7-1:User.hbm.xml 源代码片段
>
> ```xml
> <hibernate-mapping>
>     <class name=" com.xt.ass.domain .User" table="user_" lazy="true">
>         <id name="userId">
>             <generator class="native" />
>         </id>
>         <property name="username"/>
>         <property name="userpass"/>
>     </class>
> </hibernate-mapping>
> ```

通过将 class 的 lazy 属性设置为 true,来开启实体的延迟加载特性。如果我们运行示例代码 7-2 中的代码。

> 示例代码 7-2:加载 User 对象的方法
>
> ```java
> public User findUser(Integer id) {
>     Session session = null;
>     try {
>         session = HibernateSessionFactory.getSession();
>         User user = (User) session.load(User.class, id);  // (1)
>         System.out.println(user.getUsername());           // (2)
>         return user;
>     } finally {
>         if (session != null)
>             session.close();
>     }
> }
> public static void main(String[] args) {
>     UserDAO dao = new UserDAO();
>     System.out.println(dao.findUser(1).getUsername());
> }
> ```

当运行到(1)处时,Hibernate 并没有发起对数据的查询,如果我们此时通过一些调试工具(比如 JBuilder 2005 的 Debug 工具),观察此时 user 对象的内存快照,我们会惊奇地发现,此时返回的可能是 User$EnhancerByCGLIB$$bede8986 类型的对象,而且其属性为 null,这是怎么回事？前面介绍过 session.load() 方法,会返回实体对象的代理类对象,这里所返回的对象类型就是 User 对象的代理类对象。在 Hibernate 中通过使用 CGLIB,来实现动态构造一个目标对象的代理类对象,并且在代理类对象中包含目标对象的所有属性和方法,而且所有属性均被赋值为 null。通过调试器显示的内存快照,可以看出此时真正的 User 对象,是包含在代理对象的 CGLIB$CALBACK_0.target 属性中,当代码运行到（2）处时,此时调用 user.getName() 方法,这时通过 CGLIB 赋予的回调机制,实际上调用 CGLIB$CALBACK_0.getName() 方法,当调用该方法时,Hibernate 会首先检查 CGLIB$CALBACK_0.target 属性是否为 null,如果不为空,则调用目标对象的 getName() 方法,如果为空,则会发起数据库查询,生成类似这样的 SQL 语句:"select user0_.userId as userId8_0_,user0_.username as username8_0_,user0_.userpass as userpass8_0_ from user_ user0_ where user0_.userId=?"。

这样,通过一个中间代理对象,Hibernate 实现了实体的延迟加载,只有当用户真正发起获得实体对象属性的动作时,才真正会发起数据库查询操作。所以实体的延迟加载是用通过中间代理类完成的,所以只有 session.load() 方法才会利用实体延迟加载,因为只有 session.load() 方法才会返回实体类的代理类对象。

- 集合类型的延迟加载

在 Hibernate 的延迟加载机制中,针对集合类型的应用,意义是最为重大的,因为这有可能使性能得到大幅度的提高。为此,Hibernate 进行了大量的努力,其中包括对 JDK Collection 的独立实现。我们在一对多关联中定义的用来容纳关联对象的 Set 集合,并不是 java.util.Set 类型或其子类型,而是 net.sf.hibernate.collection.Set 类型,通过使用自定义集合类的实现,Hibernate 实现了集合类型的延迟加载。为了对集合类型使用延迟加载,我们必须用如示例代码 7-3 所示的代码配置我们的实体类的关于关联的部分。

```
示例代码 7-3: User 对象添加 Set 属性
<hibernate-mapping >
    <class name=" com.xt.ass.domain .User" table="user_">
        <id name="userId">
            <generator class="native" />
        </id>
        <property name="username"/>
        <property name="userpass"/>
        <set name="address" lazy="true" inverse="true">
            <key column="userId"/>
            <one-to-many class="Address" />
        </set>
    </class>
</hibernate-mapping>
```

通过将 <set> 元素的 lazy 属性设置为 true 来开启集合类型的延迟加载特性。我们看示例

代码 7-4 所示的代码。

```
示例代码 7-4：加载 User 对象
public User findUser() {
    Session session = null;
    try {
        session = HibernateSessionFactory.getSession();
        User user = (User) session.load(User.class, 1);
        Collection addset = user.getAddress(); // (1)
        Iterator it = addset.iterator(); // (2)
        while (it.hasNext()) {
            Address address = (Address) it.next();
            System.out.println(address.getAddressName());
        }
        return user;
    } finally {
        if (session != null)
            session.close();
    }
}

public static void main(String[] args) {
    UserDAO dao = new UserDAO();
    User user = dao.findUser();
    user.getAddress().size();
}
```

当程序执行到（1）处时，这时并不会发起对关联数据的查询来加载关联数据，只有运行到（2）处时，真正的数据读取操作才会开始。这时，Hibernate 会根据缓存中符合条件的数据索引，来查找符合条件的实体对象。

这里我们引入了一个全新的概念——数据索引，下面我们首先讲解一下什么是数据索引。在 Hibernate 中对集合类型进行缓存时，是分两部分进行缓存的，首先缓存集合中所有实体的 id 列表，然后缓存实体对象。这些实体对象的 id 列表，就是所谓的数据索引。当查找数据索引时，如果没有找到对应的数据索引，这时就会有一条 select SQL 的执行，获得符合条件的数据，并构造实体对象集合和数据索引，然后返回实体对象的集合，并且将实体对象和数据索引纳入 Hibernate 的缓存之中。另一方面，如果找到对应的数据索引，则从数据索引中取出 id 列表，然后根据 id 在缓存中查找对应的实体，如果找到就从缓存中返回，如果没有找到，再发起

select SQL 查询。在这里我们看出了另外一个问题，这个问题可能会对性能产生影响，这就是集合类型的缓存策略。如果我们用如示例代码 7-5 所示的代码配置集合类型。

示例代码 7-5：集合类型的缓存策略

```xml
<hibernate-mapping>
    <class name=" com.xt.ass.domain .User" table="user_">
        <id name="userId">
            <generator class="native" />
        </id>
        <property name="username"/>
        <property name="userpass"/>
        <set name="address" lazy="true" inverse="true">
            <cache usage="read-only"/>
            <key column="userId"/>
            <one-to-many class="Address" />
        </set>
    </class>
</hibernate-mapping>
```

这里我们应用了 `<cache usage="read-only"/>` 配置，如果采用这种策略来配置集合类型，Hibernate 将只会对数据索引进行缓存，而不会对集合中的实体对象进行缓存。如上配置我们运行如示例代码 7-6 所示的代码。

示例代码 7-6：加载 User 对象

```java
public User findUser() {
    Session session = null;
    try {
        session = HibernateSessionFactory.getSession();
        User user = (User) session.load(User.class, 1);
        Iterator it = user.getAddress().iterator();
        while (it.hasNext()) {
            Address address = (Address) it.next();
            System.out.println(address.getAddressName());
        }
        System.out.println("Second query……");
        User user2 = (User) session.load(User.class, 1);
        Iterator it2 = user2.getAddress().iterator();
        while (it2.hasNext()) {
            Address address2 = (Address) it2.next();
```

```
                    System.out.println(address2.getAddressName());
                }

                return user;
        } finally {
                if (session != null)
                        session.close();
        }

}

public static void main(String[] args) {
        UserDAO dao = new UserDAO();
        User user = dao.findUser();
        user.getAddress().size();
}
```

运行这段代码,会得到如图 7-1 所示的输出结果。

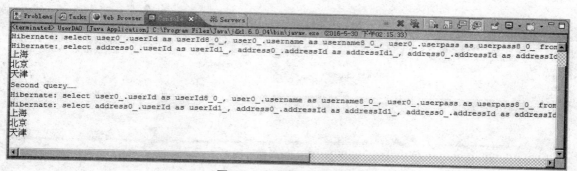

图 7-1　查询用户地址信息

我们看到,当第二次执行查询时,执行了两条对 address 表的查询操作,为什么会这样?这是因为当第一次加载实体后,根据集合类型缓存策略的配置,只对集合数据索引进行了缓存,而并没有对集合中的实体对象进行缓存。所以,在再次加载实体时,Hibernate 找到了对应实体的数据索引。但是,根据数据索引,却无法在缓存中找到对应的实体,所以 Hibernate 根据找到的数据索引发起了两条 select SQL 的查询操作。这里造成了对性能的浪费,怎样才能避免这种情况呢?我们必须对集合类型中的实体也指定缓存策略,所以我们要如示例代码 7-7 所示的代码对集合类型进行配置。

示例代码 7-7：集合类型缓存策略

```xml
<hibernate-mapping package="com.xt.ass.domain">
    <class name=" com.xt.ass.domain .User"  table="user_">
        <id name="userId">
                <generator class="native" />
        </id>
        <property name="username"/>
        <property name="userpass"/>
        <set name="address" lazy="true" inverse="true">
                <cache usage="read-write"/>
                <key column="userId"/>
                <one-to-many class="Address" />
        </set>
    </class>
</hibernate-mapping>
```

此时 Hibernate 会对集合类型中的实体也进行缓存。如果根据这个配置再次运行上面的代码，将会得到如图 7-2 所示的输出结果。

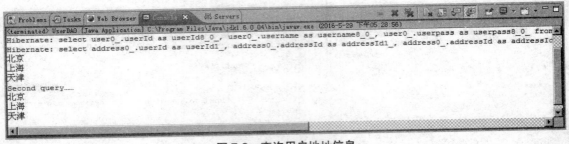

图 7-2　查询用户地址信息

这时将不会再有根据数据索引进行查询的 SQL 语句，因为此时可以直接从缓存中获得集合类型中存放的实体对象。

● 属性延迟加载

在 Hibernate 3.0 中，引入了一种新的特性——属性的延迟加载，这个机制又为获取高性能查询提供了有力的工具。假设在 User 对象中有一个 resume 字段，该字段是一个 java.sql.Clob 类型，包含了用户的简历信息，属于大数据对象。当加载用户对象时，我们不得不每一次都要加载 resume 这个字段，不论我们是否真的需要它，而且这种大数据对象的读取本身会带来很大的性能开销。在 Hibernate 2.0 中，我们只有通过我们前面讲过的面性能的粒度细分，来分解 User 类，来解决这个问题。但是，在 Hibernate 3.0 中，我们可以通过属性延迟加载机制，来使我们获得只有当我们真正需要操作这个字段时，才去读取这个字段数据的能力。为此，我们必须用如示例代码 7-8 所示的代码配置我们的实体类。

> **示例代码 7-8：属性延迟加载配置**
>
> ```xml
> <hibernate-mapping >
>     <class name=" com.xt.ass.domain .User"  table="user_ ">
>         <id name="userId">
>             <generator class="native" />
>         </id>
>         <property name="username"/>
>         <property name="userpass"/>
>         <set name="address" lazy="true" inverse="true">
>             <cache usage="read-write"/>
>             <key column="userId"/>
>             <one-to-many class="Address" />
>         </set>
>         <property type="java.sql.Clob" column="resume" lazy="true"/>
>     </class>
> </hibernate-mapping>
> ```

通过对 <property> 元素的 lazy 属性设置 true 来开启属性的延迟加载，在 Hibernate 3.0 中为了实现属性的延迟加载，使用了类增强器来对实体类的 Class 文件进行强化处理，通过增强器的增强，将 CGLIB 的回调机制逻辑，加入实体类。这里我们可以看出，属性的延迟加载还是通过 CGLIB 来实现的。CGLIB 是 Apache 的一个开源工程，这个类库可以操纵 java 类的字节码，根据字节码来动态构造符合要求的类对象。根据上面的配置我们运行示例代码 7-9 所示的代码。

> **示例代码 7-9：加载 User 对象**
>
> ```java
> public User findUser() {
>     String sql = "from User user where user.username='a' ";
>     Session session = null;
>     User user = null;
>     try {
>         session = HibernateSessionFactory.getSession();
>         Query query = session.createQuery(sql); // (1)
>         List list = query.list();
>         for (int i = 0; i < list.size(); i++) {
>             user = (User) list.get(i);
>             System.out.println(user.getUsername());
>             System.out.println(user.getResume()); // (2)
> ```

```
                        return user;
            } finally {
                   if (session != null)
                         session.close();
            }
      }

      public static void main(String[] args) {
            UserDAO dao = new UserDAO();
            User user = dao.findUser();
      }
```

当执行到（1）处时，会生成类似如下的 SQL 语句："select user0_.userId as userId8_0_, user0_.username as username8_0_, user0_.userpass as userpass8_0_ from user_ user0_ where user0_.username=?"。这时 Hibernate 会检索 User 实体中所有非延迟加载属性对应的字段数据。当执行到（2）处时，会生成类似如下的 SQL 语句："select user0_.resume as resume8_0_ from user_ user0_ where user0_.username=?"。这时会发起对 resume 字段数据真正的读取操作。

7.2　Hibernate 缓存

7.2.1　什么是 Hibernate 缓存

缓存是位于应用程序与物理数据源之间，用于临时存放复制数据的内存区域，其目的是为了减少应用程序对物理数据源访问的次数，从而提高应用程序的运行性能。Hibernate 是一个持久层框架，经常访问物理数据库。为了降低应用程序对物理数据源访问的频率，Hibernate 缓存内的数据是对物理数据源中的数据的复制，应用程序在运行时从缓存读写数据，在特定的时刻或事件会同步缓存和物理数据源的数据，从而提高应用程序的运行性能。

Hibernate 在查询数据时，首先到缓存中去查找，如果找到就直接使用，找不到的时候就会从物理数据源中检索。所以，把频繁使用的数据加载到缓存区后，就可以大大减少应用程序对物理数据源的访问，使得程序的运行性能明显提升。

7.2.2　Hibernate 缓存的分类

Hibernate 缓存包括两大类：Hibernate 一级缓存和 Hibernate 二级缓存。

Hibernate 一级缓存又称为"Session 的缓存"，它是内置的，不能被卸载（不能被卸载的意思就是这种缓存不具有可选性，必须有的功能，不可以取消 Session 缓存）。由于 Session 对象的生命周期通常对应一个数据库事务或者一个应用事务,因此它的缓存是事务范围的缓存。第一级缓存是必需的,不允许而且事实上也无法卸除。在第一级缓存中,持久化类的每个实例都具有唯一的 OID。

Hibernate 二级缓存又称为"SessionFactory 的缓存"。由于 SessionFactory 对象的生命周期和应用程序的整个过程对应,因此 Hibernate 二级缓存是进程范围或者集群范围的缓存,有可能出现并发问题。因此,需要采用适当的并发访问策略,该策略为被缓存的数据提供了事务隔离级别。第二级缓存是可选的,是一个可配置的插件,在默认情况下,SessionFactory 不会启用这个插件。

SessionFactory 的缓存分为内置缓存和外置缓存。内置缓存中存放的是 SessionFactory 对象的一些集合属性包含的数据(映射元素据及预定义 SQL 语句等),对于应用程序来说,它是只读的。外置缓存中存放的是数据库数据的副本,其作用和一级缓存类似。二级缓存除了以内存作为存储介质外,还可以选用硬盘等外部存储设备。

7.2.3 Hibernate 的缓存范围

Hibernate 的一级缓存和二级缓存均位于持久层,且均用于存放数据库数据的副本,最大的区别就是缓存的范围各不相同。

缓存的范围分为三类:

- 事务范围

事务范围的缓存只能被当前事务访问。每个事务都有各自的缓存,缓存内的数据通常采用相互关联的对象形式。缓存的生命周期依赖于事务的生命周期,只有当事务结束时,缓存的生命周期才会结束。事务范围的缓存使用内存作为存储介质,一级缓存就属于事务范围。

- 应用范围

应用程序的缓存可以被应用范围内的所有事务共享访问。缓存的生命周期依赖于应用的生命周期,只有当应用结束时,缓存的生命周期才会结束。应用范围的缓存可以使用内存或硬盘作为存储介质,二级缓存就属于应用范围。

- 集群范围

在集群环境中,缓存被一个机器或多个机器的进程共享,缓存中的数据被复制到集群环境中的每个进程节点,进程间通过远程通信来保证缓存中的数据的一致,缓存中的数据通常采用对象的松散数据形式。

7.2.4 Hibernate 的缓存管理

- 一级缓存的管理

Session 中提供了如下几个方法对一级缓存进行管理:

evict(Object obj):将指定的持久化对象从一级缓存中清除,释放对象所占用的内存资源,指定对象从持久化状态变为脱管状态,从而成为游离对象。

clear():将一级缓存中的所有持久化对象清除,释放其占用的内存资源。

contains(Object obj):判断指定的对象是否存在于一级缓存中。

flush():刷新一级缓存区的内容,使之与数据库数据保持同步。

● 二级缓存的管理

SessionFactory 中提供了如下方法对二级缓存进行管理：

evict(Class arg0, Serializable arg1)：将某个类的指定 ID 的持久化对象从二级缓存中清除，释放对象所占用的资源。

例如：

```
sessionFactory.evict(Customer.class, new Integer(1));
sessionFactory.evict(Customer.class, new Integer(1));
```

evict(Class arg0)：将指定类的所有持久化对象从二级缓存中清除，释放其占用的内存资源。

例如：

```
sessionFactory.evict(Customer.class);
sessionFactory.evict(Customer.class);
```

evictCollection(String arg0)：将指定类的所有持久化对象的指定集合从二级缓存中清除，释放其占用的内存资源。

例如：

```
sessionFactory.evictCollection("Customer.orders");
sessionFactory.evictCollection("Customer.orders");
```

7.2.5　Hibernate 的二级缓存的配置

首先，不是所有的数据都适合放在二级缓存中。那么，什么样的数据适合放在二级缓存中，什么样的数据不适合放在二级缓存中呢？

下面这几种情况就不适合加载到二级缓存中：

（1）经常被修改的数据。
（2）绝对不允许出现能并发访问的数据。
（3）与其他应用共享的数据。

下面这几种情况合适加载到二级缓存中：

（1）数据更新频率低。
（2）允许偶尔出现并发问题的非重要数据。
（3）不会被并发访问的数据。
（4）常量数据。
（5）不会被第三方修改的数据。

Hibernate 的二级缓存功能是靠配置二级缓存插件来实现的。为了集成这些插件，Hibernate 提供了 org.hibernate.cache.CacheProvider 接口，它充当缓存插件与 Hibernate 之间的适配器。

常用的二级缓存插件包括以下几种：

第7章 Hibernate 高级特性

```
EHCache    org.hibernate.cache.EhCacheProvider
OSCache    org.hibernate.cache.OSCacheProvider
SwarmCahe  org.hibernate.cache.SwarmCacheProvider
JBossCache org.hibernate.cache.TreeCacheProvider
```

本节，我们简单介绍一下 EHCache 的配置，其他缓存插件的配置模式基本类似，这里不再赘述。

第一步，在 hibernate 配置文件中设置二级缓存，代码如示例代码 7-10 所示。

示例代码 7-10：hibernate.cfg.xml 配置文件源代码片段

```xml
<hibernate-configuration>
  <session-factory>
    <property name="connection.password">123456</property>
    <property name="connection.username">sa</property>
    <property name="connection.url">jdbc:sqlserver://127.0.0.1:1433;databaseName=advance</property>
    <property name="dialect">org.hibernate.dialect.SQLServerDialect</property>
    <property name="myeclipse.connection.profile">db1</property>
    <property name="connection.driver_class">com.microsoft.sqlserver.jdbc.SQLServerDriver</property>
    <property name="show_sql">true</property>
    <!-- 设置二级缓存插件 EHCache 的 Provider 类 -->
    <property name="hibernate.cache.provider_class">
      org.hibernate.cache.EhCacheProvider
    </property>
    <!-- 启动 "查询缓存" -->
    <property name="hibernate.cache.use_query_cache">
      true
    </property>
    <!-- 省略部分代码 -->
  </session-factory>
</hibernate-configuration>
```

第二步，新建缓存配置文件，命名为 ehcache.xml，代码如示例代码 7-11 所示。

示例代码 7-11：ehcache.xml 配置文件源代码片段

```xml
<ehcache>
 <!--
 maxElementsInMemory 为缓存对象的最大数目,
 eternal 设置是否永远不过期,timeToIdleSeconds 对象处于空闲状态的最多秒数,
 timeToLiveSeconds 对象处于缓存状态的最多秒数
 -->
  <diskStore path="java.io.tmpdir"/>
    <defaultCache  maxElementsInMemory="10000" eternal="false"  timeToIdleSeconds="300" timeToLiveSeconds="600" overflowToDisk="true"/>
</ehcache>
```

第三步，在指定的映射文件中配置缓存信息，例如在 User.hbm.xml 中添加如示例代码 7-12 所示的配置。

示例代码 7-12：User.hbm.xml 配置文件源代码片段

```xml
<?xml version="1.0" encoding='UTF-8'?>
<!DOCTYPE hibernate-mapping PUBLIC
"-//Hibernate/Hibernate Mapping DTD 3.0//EN"
"http://hibernate.sourceforge.net/hibernate-mapping-3.0.dtd" >

<hibernate-mapping>
  <class>
      <!-- 设置该持久化类的二级缓存并发访问策略,可选的有:
read-only、read-write、nonstrict-read-write、transactional
-->
      <cache usage="read-write"/>
      <id name="userId">
              <generator class="native" />
          </id>
      <property name="username"/>
      <property name="userpass"/>
      <set name="address" lazy="true" inverse="true">
              <cache usage="read-write"/>
              <key column="userId"/>
              <one-to-many class="Address" />
          </set>
  </class>
</hibernate-mapping>
```

● Hibernate 的二级缓存策略的一般过程如下：

（1）条件查询的时候，总是发出一条类似"select * from table_name where ……"（选择所有字段）这样的 SQL 语句查询数据库，一次获得所有的数据对象。

（2）把获得的所有数据对象根据 ID 放入第二级缓存中。

（3）当 Hibernate 根据 ID 访问数据对象的时候，首先从 Session 一级缓存中查；查不到，如果配置了二级缓存，那么从二级缓存中查；查不到，再查询数据库，把结果按照 ID 放入到缓存。

（4）删除、更新、增加数据的时候，同时更新缓存。

Hibernate 的二级缓存策略，是针对于 ID 查询的缓存策略，对于条件查询则毫无作用。为此，Hibernate 提供了针对条件查询的 Query 缓存。

● Hibernate 的 Query 缓存策略的过程如下：

（1）Hibernate 首先根据这些信息组成一个 Query Key，Query Key 包括条件查询一般信息：SQL，SQL 需要的参数，记录范围（起始位置 rowStart，最大记录个数 maxRows）等。

（2）Hibernate 根据这个 Query Key 到 Query 缓存中查找对应的结果列表。如果存在，那么返回这个结果列表；如果不存在，查询数据库，获取结果列表，把整个结果列表根据 Query Key 放入到 Query 缓存中。

（3）Query Key 中的 SQL 涉及一些表名，如果这些表的任何数据发生修改、删除、增加等操作，这些相关的 Query Key 都要从缓存中清空。

简单地说，当 Hibernate 根据 ID 访问数据对象的时候，首先从 Session 一级缓存中查，如果查不到，则检查是否配置了二级缓存。如果配置了二级缓存，那么从二级缓存中查。如果两级缓存都查不到，再查询数据库，把结果按照 ID 放入缓存中。删除、更新、增加数据的时候，同时更新缓存。

无论何时，我们在管理 Hibernate 缓存（Managing the caches）时，当你给 save()、update() 或 saveOrUpdate() 方法传递一个对象时，或使用 load()、get()、list()、iterate() 或 scroll() 方法获得一个对象时，该对象都将被加入到 Session 的内部缓存中。

当随后 flush() 方法被调用时，对象的状态会和数据库取得同步。如果你不希望此同步操作发生，或者你正处理大量对象，需要有效管理内存时，你可以调用 evict() 方法，从一级缓存中去掉这些对象及其集合。

缓存提前访问的好处是：表现层的请求不必经过业务层直接得到缓存，节省业务层开销。缓存越靠近用户界面端，性能越好，查询越频繁使用，缓存击中率越高，各方面消耗就越小。最好缓存在客户端，这也是胖客户端一个性能优势。

至此，我们对高级特性有了更深刻的理解，Hibernate（冬眠）是发生在冬天的故事，从下一章我们将开始学习另一个神奇的框架技术 Spring（春天），我们将在 Spring 的绿草丛中尽情感受程序设计是一件多么激动人心的事情。

7.3 小结

✓ Hibernate 的检索策略包括类级别检索策略和关联级别检索策略。

- ✓ 在 Hibernate 中提供了对实体对象的延迟加载以及对集合的延迟加载,另外在 Hibernate 3.0 中还提供了对属性的延迟加载。
- ✓ 缓存是位于应用程序与物理数据源之间,用于临时存放复制数据的内存区域。
- ✓ Hibernate 缓存包括两大类:Hibernate 一级缓存和 Hibernate 二级缓存。

7.4　英语角

lazy	延迟
inverse	反转
cache	缓存
usage	策略

7.5　作业

1. 在房屋租赁系统中,通过设置 Fwxx 映射文件,使得 JD 集合属性延迟加载。
2. 在房屋租赁系统中,通过添加二级缓存插件并设置 Fwxx 映射文件,使得 Fwxx 对象能够存放于二级缓存中。

7.6　思考题

Hibernate 的一级缓存和二级缓存有什么区别?

7.7　学员回顾内容

1. Hibernate 检索策略。
2. Hibernate 延迟加载。
3. Hibernate 缓存机制。

第 8 章 Spring 原理与应用

学习目标

- ◆ 了解 Spring 的发展。
- ◆ 了解 Spring 的基本组成。
- ◆ 掌握 Spring 环境搭建。
- ◆ 掌握实例化 bean 的三种方法。
- ◆ 了解 bean 的作用域范围。

课前准备

通过网络或其他途径了解什么是面向接口编程。

本章简介

前面已经学习了 Struts 框架和 Hibernate 框架，从本章开始学习 Spring 框架。Spring 到今天已经发展为一个功能丰富而且易用的集成框架，但其最重要的特点是"依赖注入"（Dependency Injection，DI）模式和"面向方面编程"（Aspect-Oriented programming，AOP）原理的优秀实现。

通过前面的学习我们认识到，框架可以帮助我们构建规范的、优秀的应用程序，还会简化繁琐的编码过程，Spring 的作用两者兼而有之。Spring 的目标就是"使 Java EE 更易用并促进好的编程习惯"。

Struts 是第一个开源的 Java Web 框架，给我们提供了优秀的 MVC 支持；Hibernate 大大简化了持久化的代码。Spring 更大意义上充当了黏合剂和润滑剂的角色，它对 Struts、Hibernate 等技术都提供了良好的支持，能够把现有的 Java 系统柔顺地整合起来，并让它们更易用。同时自身还提供了声明式事务等企业级开发不可或缺的功能。

在 EJB 盛行于世的时代，J2EE（Java EE 之前称作 J2EE）变得庞大和沉重，开发人员在严冬中坚忍前行。2003 年春天，Spring 降临世间，从此，Java 世界改变了。

8.1 Spring 概述

8.1.1 Spring 创造人——Rod Johoson

Rod Johnson 毕业于悉尼大学，他不仅获得了计算机学位，同时还获得了音乐学博士学位。有着相当丰富的 C/C++ 技术背景。早在 1996 年，他就开始了对 Java 服务器端技术的研究（Java 在 1995 年才正名为 Java）。他是一个在保险、电子商务和金融行业有着丰富经验的技术顾问，同时也是 JSR—154（Servlet 2.4）和 JDO 2.0 的规范专家、JCP（讨论了制定 Java 标准的组织）的积极成员。图 8-1 所示即为 Rod Johoson 的画像。

图 8-1　Rod Johoson

2002 年他出版了《Expert One-on-One J2EE 设计与开发》。在书中，他对传统的 J2EE 技术（以 EJB 为核心）日益臃肿和低效提出了质疑，他觉得应该有更便捷的做法，于是提出了 Interface21，也就是 Spring 框架的雏形。他提出了技术以实用为准的主张，引发了人们对"正统"J2EE 的反思。2003 年 2 月他创建了 Spring 开源项目，开办了 Interface21 公司，专门提供 SpringFramework 培训和咨询，任 CEO，现居伦敦，和 Gavin King 的风格不同，Mr Johnson 是一位典型绅士，讲话一板一眼字正腔圆。

2005 年的 the server side symposium 上，JBoss 那帮家伙戴着面具穿着小丑衣（扮成蝙蝠侠里 Joker 的是大名鼎鼎的 MarcFleury，JBoss 的创始人，如图 8-2 所示）出席会议时，Rod 说他自己和整个 Spring 项目组都没啥幽默感，也不会穿那种衣服出席，结果引来了 Gavin King 以他一贯的富有激情而又节制的风格在自己的 Blog 上做出了尖锐的回应。虽然个性迥异，但有一点是一样的，他们都在日复一日地为 Java 社区默默地做出伟大的贡献，为人类创造了巨大价值。光荣在于平淡，卓越在于专注。

图 8-2 MarcFleury

8.1.2 Spring 的绿草丛

Spring 确实给人一种格外清新、爽朗的感觉。仿佛微雨后的绿草丛，讨人喜欢，又蕴藏着勃勃生机。它大大简化了 Java 企业级开发，提供了强大、稳定的功能，又没有带来额外的负担。让人们使用 Spring 做每一件事情的时候都有得体和优雅的感觉。Spring 有两个主要目标：一是让现有技术更易于使用，二是促进良好的编程习惯（或者称为最佳实践）。

Spring 是一个全面的解决方案。但它坚持一个原则：不重新造轮子。已经有较好解决方案的领域，Spring 绝不做重复性的实现，比如对象持久化和 OR 映射，Spring 只是对现有 JDBC、Hibernate、JPA 等技术提供支持，使之更易用，而不是重新做一个实现。

Spring Framework 图标如图 8-3 所示。

图 8-3 Spring 框架的 logo

Spring 依然在不断发展和完善，但基本与核心的部分已经相当稳定，包括 Spring 的依赖注入容器、AOP 实现和对持久化层的支持。本章将介绍依赖注入，后面的章节我们介绍 AOP 和 Spring 对持久化层的支持以及 AOP 的一个应用：声明式事务。

图 8-4 描述了 Spring Framework 包含的内容。其中最基础的是 Spring Core，即 Spring 作为依赖注入容器的部分。Spring AOP 是基于 Spring Core 实现的，是一个典型的声明式事务。Spring DAO 对 JDBC 提供了支持，简化了 JDBC 编码，同时使代码更健壮。Spring ORM 部分对 Hibernate 等 OR 映射框架提供了支持。Spring 可以在 Java SE 中使用，也可以在 Java EE 中使用，Spring Context 为企业级开发提供了便利和集成的工具。Spring Web 是为 Spring 在 Web 应用程序中使用提供的支持，在后面的章节我们将详细介绍。

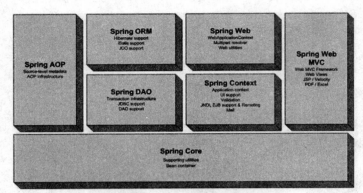

图 8-4 Spring Framework 组成

我们知道 Struts 是基于 MVC 的,但编码略显笨拙,Spring 提供了一个稍微简单的 Spring Web MVC 框架,但应用不多。

组成 Spring 框架的每个模块(或组件)都可以单独存在,或者与其他一个或多个模块联合实现。每个模块的功能如下:

● 核心容器:核心容器提供 Spring 框架的基本功能。核心容器的主要组件是 BeanFactory,它是工厂模式的实现。BeanFactory 使用控制反转(IOC)模式将应用程序的配置和依赖性规范与实际的应用程序代码分开。

● Spring 上下文:Spring 上下文是一个配置文件,向 Spring 框架提供上下文信息。Spring 上下文包括企业服务,例如 JNDI、EJB、电子邮件、国际化、校验和调度功能。

● Spring AOP:通过配置管理特性,Spring AOP 模块直接将面向方面的编程功能集成到了 Spring 框架中。所以,可以很容易地使 Spring 框架管理的任何对象支持 AOP。Spring AOP 模块为基于 Spring 的应用程序中的对象提供了事务管理服务。通过使用 Spring AOP,不用依赖 EJB 组件,就可以将声明性事务管理集成到应用程序中。

● Spring DAO:JDBC DAO 抽象层提供了有意义的异常层次结构,可用该结构来管理异常处理和不同数据库供应商抛出的错误消息。异常层次结构简化了错误处理,并且极大地降低了需要编写的异常代码数量(例如打开和关闭连接)。Spring DAO 的面向 JDBC 的异常遵从通用的 DAO 异常层次结构。

● Spring ORM:Spring 框架插入了若干个 ORM 框架,从而提供了 ORM 的对象关系工具,其中包括 JDO、Hibernate 和 iBatis SQL Map。这些都遵从 Spring 的通用事务和 DAO 异常层次结构。

● Spring Web 模块:Web 上下文模块建立在应用程序上下文模块之上,为基于 Web 的应用程序提供了上下文。所以,Spring 框架支持与 Jakarta Struts 的集成。Web 模块还简化了处理多部分请求以及将请求参数绑定到域对象的工作。

● Spring MVC 框架:MVC 框架是一个全功能的构建 Web 应用程序的 MVC 实现。通过策略接口,MVC 框架变成高度可配置的。MVC 容纳了大量视图技术,其中包括 JSP、Velocity、Tiles、iText 和 POI。

Spring 框架的功能可以用在任何 J2EE 服务器中,大多数功能也适用于不受管理的环境。Spring 的核心要点是:支持不绑定到特定 J2EE 服务的可重用业务和数据访问对象。毫无疑问,这样的对象可以在不同 J2EE 环境(Web 或 EJB)、独立应用程序、测试环境之间重用。

8.1.3 Spring 的优点

Spring 是一个开源框架,是为了解决企业应用程序开发复杂性而创建的。框架的主要优势之一就是其分层架构,分层架构允许您选择使用哪一个组件,同时为 J2EE 应用程序开发提供集成的框架。总结起来,Spring 具有以下优点:
- 提供了一个一致的编程模型。
- 旨在促进代码重用。
- 旨在促进面向对象的设计。
- 致力于推动好的编程习惯,如用接口编程。
- Spring 改进了从 Java 代码中提取配置值到 XML 或者属性文件中的方法。
- 在项目中引入 Spring 可以带来下面的好处:降低组件之间的耦合度,实现软件各层之间的解耦。参见图 8-5。

图 8-5 组件间解耦

- 可以使用容器提供的众多服务,如:事务管理服务、消息服务等。当我们使用容器管理事务时,开发人员就不再需要手工控制事务,也不需处理复杂的事务传播。
- 容器提供单例模式支持,开发人员不再需要自己编写实现代码。
- 容器提供了 AOP 技术,利用它很容易实现如权限拦截、运行期监控等功能。
- 容器提供了众多辅助类,使用这些类能够加快应用的开发,如:JdbcTemplate、HibernateTemplate。
- Spring 对于主流的应用框架提供了集成支持,如:集成 Hibernate、JPA、Struts 等,这样更便于应用的开发。

8.2 Spring 核心技术

Spring 是一个开源的控制反转(Inversion of Control, IoC)和面向方面(AOP)的容器框架。它的主要目的是简化企业开发。
- 控制反转(Inversion of Control, IoC)即好莱坞原则:"Don't call me, I'll call you."

IoC 的意义就是一种依赖关系的转移,IoC 要求的是:
✧ 容器不应该侵入应用程序,就是不应出现与容器相依的 API。
✧ 应用程序本身可以依赖于抽象接口,容器透过这些抽象接口注入需要的资源至应用程序中。

IoC 的实现方式:
✧ Dependency Injection(Spring 所采用)。

◇ Service Locater。
- 依赖注入（Dependency Injection，DI）：是 IoC 的具体实现，DI 的三种实现方式：
 ◇ Interface Injection（Type 1）。
 ◇ Setter Injection（Type 2）。
 ◇ Constructor Injection（Type 3）。
- AOP（Aspect-Oriented Programming）

面向方面编程给 POJO 对象提供声明式的服务，而该对象无须实现特殊的 API，将提供（或插入）的服务独立出来设计成对象，这样的对象称为 Aspect。AOP 注重以下几个方面的管理：

 ◇ 事务管理。
 ◇ 日志记录。
 ◇ 故障监控。
 ◇ 测试。

8.2.1 IOC

控制反转模式（也称作依赖性介入）的基本概念是：不创建对象，但是描述创建它们的方式。在代码中不直接与对象和服务连接，但在配置文件中描述哪一个组件需要哪一项服务。容器（在 Spring 框架中是 IOC 容器）负责将这些联系在一起。

在典型的 IOC 场景中，容器创建了所有对象，并设置必要的属性将它们连接在一起，决定什么时间调用方法。下表列出了 IOC 的一个实现模式。

表 8-1 IOC 的实现模式

类型	实现模式
类型 1	服务需要实现专门的接口，通过接口，由对象提供这些服务，可以从对象查询依赖性（例如，需要的附加服务）
类型 2	通过 JavaBean 的属性（例如 setter 方法）分配依赖性
类型 3	依赖性以构造函数的形式提供，不以 JavaBean 属性的形式公开

Spring 框架的 IOC 容器采用类型 2 和类型 3 实现。

8.2.2 AOP

面向方面的编程，即 AOP，是一种编程技术，它允许程序员对横切关注点或横切典型的职责分界线的行为（例如日志和事务管理）进行模块化。AOP 的核心构造是方面，它将那些影响多个类的行为封装到可重用的模块中。

AOP 和 IOC 是补充性的技术，它们都运用模块化方式解决企业应用程序开发中的复杂问题。在典型的面向对象开发方式中，可能要将日志记录语句放在所有方法和 Java 类中才能实现日志功能。在 AOP 方式中，可以反过来将日志服务模块化，并以声明的方式将它们应用到需要日志的组件上。当然，优势就是 Java 类不需要知道日志服务的存在，也不需要考虑相关的代码。所以，用 Spring AOP 编写的应用程序代码是松散耦合的。

AOP 的功能完全集成到了 Spring 事务管理、日志和其他各种特性的上下文中。

8.2.3 IOC 容器

Spring 设计的核心是 org.springframework.beans 包,它的设计目标是与 JavaBean 组件一起使用。这个包通常不是由用户直接使用,而是由服务器将其用作其他多数功能的底层中介。下一个最高级抽象是 BeanFactory 接口,它是工厂设计模式的实现,允许通过名称创建和检索对象。BeanFactory 也可以管理对象之间的关系。

BeanFactory 支持两个对象模型:
- 单态:模型提供了具有特定名称的对象的共享实例,可以在查询时对其进行检索。Singleton 是默认的也是最常用的对象模型,对于无状态服务对象很理想。
- 原型:模型确保每次检索都会创建单独的对象。当每个用户都需要自己的对象时,原型模型最适合。

bean 工厂的概念是 Spring 作为 IOC 容器的基础。IOC 将处理事情的责任从应用程序代码转移到框架。正如我将在下一个示例中演示的那样,Spring 框架使用 JavaBean 属性和配置数据来指出必须设置的依赖关系。

8.2.4 Spring 环境搭建

Spring 环境搭建和前面所介绍的各类框架搭建方式类似,无外乎还是构建环境路径,添加配置文件等,具体步骤如下:

(1)导入使用 Spring 所需要的 jar 包

我们可以到 http://www.springsource.org/download 下载 Spring 框架的源代码,然后进行解压缩,在解压目录中找到下面 jar 文件,拷贝到类路径下,主要的包如下:
- dist\spring.jar
- lib\jakarta-commons\commons-logging.jar

如果使用了面向方面编程(AOP),还需要下列 jar 文件:
- lib/aspectj/aspectjweaver.jar 和 aspectjrt.jar
- lib/cglib/cglib-nodep-2.1_3.jar

如果使用了 JSR-250 中的注解,如 @Resource/@PostConstruct/@PreDestroy,还需要下列 jar 文件:
- lib\j2ee\common-annotations.jar

(2)在项目类路径下添加 Spring 的配置文件

模版如下:

```xml
<?xml version="1.0" encoding="UTF-8"?>
<beans
    xmlns="http://www.springframework.org/schema/beans"
    xmlns:xsi="http://www.w3.org/2001/XMLSchema-instance"
    xmlns:p="http://www.springframework.org/schema/p"
    xsi:schemaLocation="http://www.springframework.org/schema/beans    http://www.springframework.org/schema/beans/spring-beans-3.1.xsd">
</beans>
```

该配置模版可以从 Spring 的参考手册或 Spring 的例子中得到。配置文件的取名可以任意，默认名称是"applicationContext.xml"。文件可以存放在任何目录下，但考虑到通用性，一般放在类路径下。

小贴士

> 由于 Spring 的 schema 文件位于网络上，如果机器不能连接到网络，那么在编写配置信息时就无法出现提示信息，解决方法有两种：
> 1. 让机器上网，MyEclipse 会自动从网络上下载 schema 文件并缓存在硬盘上。
> 2. 手动添加 schema 文件，方法如下：
> 在菜单栏中选择 windows → preferences → myeclipse → files and editors → xml → xmlcatalog
> 点击"add"按钮，在出现的窗口中的 Key Type 中选择 URI，在 location 中选"File System"，然后在 Spring 解压目录的 dist/resources 目录中选择 spring-beans-3.1.xsd，回到设置窗口的时候不要急着关闭窗口，应把窗口中的 Key Type 改为 Schema location，Key 改为"http://www.springframework.org/schema/beans/spring-beans-3.1.xsd"。

（3）实例化 Spring 容器

实例化 Spring 容器常用的两种方式：

方法一：

在类路径下寻找配置文件来实例化容器

```
ApplicationContext ctx = new ClassPathXmlApplicationContext(new String[]{"applicationContext.xml"});
```

方法二：

在文件系统路径下寻找配置文件来实例化容器

```
ApplicationContext ctx = new FileSystemXmlApplicationContext(new String[]{"d:\\applicationContext.xml"});
```
注意：Spring 的配置文件可以指定多个，可以通过 String 数组传入。

（4）从 Spring 容器中得到 bean 实例

当 Spring 容器启动后，因为 Spring 容器可以管理 bean 对象的创建、销毁生命周期等，所以我们只需从容器直接获取 bean 对象就行，而不用编写一句代码来创建 bean 对象。从容器获取 bean 对象的代码如下：

```
ApplicationContext ctx = new ClassPathXmlApplicationContext("applicationContext.xml");
PersonService service = (PersonService)ctx.getBean("personService");
```

这样，一个简单的 Spring 框架就搭建起来并可以使用了。

8.3 Spring 中的 Bean

8.3.1 bean 的实例化

在 Spring 中，bean 的实例化方式有三种，如下所示：
- 使用类构造器实例化

```
<bean id="personService " class="com.xtgj. j2ee.chapter08.bean.PersonServiceBean"/>
```

- 使用静态工厂方法实例化

```
<bean id="personService" class="com.xtgj. j2ee.chapter08.service.PersonFactory" factory-method=" createPersonService "/>
```

上述实例化方法中用到的静态工厂类源代码片段如下：

```
public class PersonFactory {
    public static PersonServiceBean createPersonService (){
        return new PersonServiceBean();
    }
}
```

- 使用实例工厂方法实例化

```
<bean id="personServiceFactory" class="com.xtgj. j2ee.chapter08.service.PersonServiceFactory "/>
<bean id="personService" factory-bean="personServiceFactory" factory-method=" createPersonService "/>
```

上述实例化方法中用到的实例工厂类源代码片段如下：

```
public class PersonServiceFactory {
    public PersonServiceBean createPersonService (){
            return new PersonServiceBean();
    }
}
```

8.3.2 Bean 的作用域

在 Spring 2.0 之前 bean 只有 2 种作用域：singleton、non-singleton(prototype)，Spring 2.0 以后增加了 session、request、global session 三种专用于 Web 应用程序上下文的 bean。因此默认情况下 Spring 现在有五种类型的 bean。

（1）singleton 作用域

当一个 bean 的作用域设置为 singleton，那么 Spring IOC 容器中只会存在一个共享的 bean 实例，并且所有对 bean 的请求，只要 id 与该 bean 定义匹配，则只会返回 bean 的同一实例。即 Spring IOC 容器只会创建该 bean 定义的唯一实例，这个单一实例会被缓存到单例缓存（singleton cache）中，并且所有针对该 bean 的后续请求和引用都将返回被缓存的对象实例。

配置实例：

```
<bean id="role" class="com.xtgj.j2ee.chapter08.bean.Role" scope="singleton"/>
或
<bean id="role" class="com.xtgj.j2ee.chapter08.bean.Role" singleton="true"/>
```

（2）prototype 作用域

prototype 作用域的 bean 会导致每次请求（将其注入另一个 bean 中，或者以程序的方式调用容器的 getBean() 方法）都会产生一个新的 bean 实例，相当于一个 new 的操作。对于 prototype 作用域的 bean，有一点非常重要，那就是 Spring 不能对一个 prototype bean 的整个周期负责。容器在初始化、配置、装饰或者是装配完一个 prototype 实例后，将它交给客户端，随后就对该 prototype 实例不闻不问了。不管何种作用域，容器都会调用所有对象的初始化生命周期回调方法，而对 prototype 而言，任何配置好的析构生命周期回调方法都不会被调用。清除 prototype 作用域的对象并释放任何 prototype bean 所持有的昂贵资源，都是客户端代码的职责（让 Spring 容器释放被 singleton 作用域 bean 占用资源的一种可行方式是，通过使用 bean 的后置处理器，该处理器持有要被清除的 bean 的引用）。

配置实例：

```
<bean id="role" class="com.xtgj.j2ee.chapter08.bean.Role" scope="prototype"/>
或
<beanid="role" class="com.xtgj.j2ee.chapter08.bean.Role" singleton="false"/>
```

（3）request 作用域

request 表示针对每一次 HTTP 请求都会产生一个新的 bean，同时该 bean 仅在当前 HTTP request 内有效。

request、session、global session 使用的时候,首先要在初始化 Web 应用的 web.xml 文件中做如下配置(如果你使用的是 Servlet 2.4 及以上的 Web 容器,那么你仅需要在 Web 应用的 web.xml 文件中增加下述 ContextListener 即可):

配置实例:

```
<web-app>
...
<listener>
<listener-class>org.springframework.web.context.request.RequestContextListener</listener-class>
</listener>
...
</web-app>
```

(4) session 作用域

session 作用域表示针对每一次 HTTP 请求都会产生一个新的 bean,同时该 bean 仅在当前 HTTP session 内有效。和 request 配置实例的前提一样,先配置好 Web 启动文件,然后在 Spring 配置文件中添加如下配置:

```
<bean id="role" class="com.xtgj.j2ee.chapter08.bean.Role" scope="session"/>
<bean id="role" class="com.xtgj.j2ee.chapter08.bean.Role" scope="session"/>
```

(5) global session 作用域

global session 作用域类似于标准的 HTTP Session 作用域,不过它仅仅在基于 portlet 的 Web 应用中才有意义。portlet 规范定义了全局 Session 的概念,它被所有构成某个 portlet Web 应用的各种不同的 portlet 所共享。在 global session 作用域中定义的 bean 被限定于全局 portlet Session 的生命周期范围内。如果你在 Web 中使用 global session 作用域来标识 bean,那么,Web 会自动当成 session 类型来使用。和 request 配置实例的前提一样,先配置好 Web 启动文件就,然后在 Spring 配置文件中添加如下配置:

```
<bean id="role" class="com.xtgj.j2ee.chapter08.bean.Role" scope="global session"/>
<bean id="role" class="com.xtgj.j2ee.chapter08.bean.Role" scope="global session"/>
```

(6) 自定义 bean 装配作用域

在 Spring 2.0 中作用域是可以任意扩展的,你可以自定义作用域,甚至你也可以重新定义已有的作用域(但是你不能覆盖 singleton 和 prototype)。Spring 的作用域由接口 org.springframework.beans.factory.config.Scope 来定义,自定义自己的作用域只要实现该接口即可。

8.3.3 指定 Bean 的初始化方法和销毁方法

指定 Bean 的初始化方法和销毁方法,可以通过如下配置实现:

```xml
<bean id="xxx" class=" com.xtgj.j2ee.chapter08.bean.PersonServiceBean" init-method="init" destroy-method="close"/>
```

Spring 提供了几个标志接口（marker interface），这些接口用来改变容器中 bean 的行为；它们包括 InitializingBean 和 DisposableBean。

实现这两个接口的 bean 在初始化和析构时容器会调用前者的 afterPropertiesSet() 方法，以及后者的 destroy() 方法。

Spring 在内部使用 BeanPostProcessor 实现来处理它能找到的任何标志接口并调用相应的方法。如果需要自定义特性或者生命周期行为，你可以实现自己的 BeanPostProcessor。

初始化回调和析构回调：

在 PersonServiceBean 中添加 init() 和 destroy() 方法表示初始化方法和销毁方法，源代码如下：

```java
package com.xtgj.j2ee.chapter08.bean;

import com.xtgj.j2ee.chapter08.service.PersonService;

public class PersonServiceBean implements PersonService{
    // 构造器
    public PersonServiceBean(){
        System.out.println("instance me");
    }

    //save 方法
    public void save() {
        System.out.println("save");
    }

    // 初始化方法,这个方法是类被实例化了之后就会执行的!
    public void init(){
        System.out.println("init");
    }

    // 销毁方法
    public void destroy(){
        System.out.println("destroy");
    }
}
```

编写 JUnit 测试类 SpringTest.java，源代码如下：

```java
package com.xtgj.j2ee.chapter08.test;

import org.junit.BeforeClass;
import org.junit.Test;
import org.springframework.context.support.AbstractApplicationContext;
import org.springframework.context.support.ClassPathXmlApplicationContext;

import com.xtgj.j2ee.chapter08.service.PersonService;

public class SpringTest {

    @BeforeClass
    public static void setUpBeforeClass() throws Exception {
    }

    @Test  // 创建的单元测试
    public void testSave() {
        AbstractApplicationContext ctx = new ClassPathXmlApplicationContext("applicationContext.xml");
        PersonService ps = (PersonService)ctx.getBean("personService");
        ps.save();
        ctx.close(); // 有了这一句才会有 destroy 方法的调用
    }

}
```

运行测试，得到如下执行结果：

第一种情况，如果 lazy-init（默认是 default，也就是 false）没有设置，或者设置为 default 或者 false，即：

```xml
<?xml version="1.0" encoding="UTF-8"?>
<beans
    xmlns="http://www.springframework.org/schema/beans"
    xmlns:xsi="http://www.w3.org/2001/XMLSchema-instance"
    xmlns:p="http://www.springframework.org/schema/p"
    xsi:schemaLocation="http://www.springframework.org/schema/beans    http://www.springframework.org/schema/beans/spring-beans-3.1.xsd">

    <bean id="personService" class="com.xtgj.j2ee.chapter08.bean.PersonServiceBean"
        scope="singleton" init-method="init" destroy-method="destroy" lazy-init="false">
    </bean>

</beans>
```

那么控制台上显示的结果是：

```
log4j:WARN No appenders could be found for logger (org.springframework.core.env.StandardEnvironment).
log4j:WARN Please initialize the log4j system properly.
log4j:WARN See http://logging.apache.org/log4j/1.2/faq.html#noconfig for more info.
instance me
init
save
destroy
```

反之，如果设置为 true，控制台上显示的结果是：

```
log4j:WARN No appenders could be found for logger (org.springframework.core.env.StandardEnvironment).
log4j:WARN Please initialize the log4j system properly.
log4j:WARN See http://logging.apache.org/log4j/1.2/faq.html#noconfig for more info.
instance me
init
save
destroy
```

修改测试代码为：

```
@Test
public void testSave() {
    AbstractApplicationContext ctx = new ClassPathXmlApplicationContext("applicationContext.xml ");
        PersonService ps = (PersonService)ctx.getBean("personService");
        PersonService ps1 = (PersonService)ctx.getBean("personService");
        System.out.println(ps==ps1);
        ctx.close();
}
```

重新测试,控制台上显示的结果为:

```
log4j:WARN No appenders could be found for logger (org.springframework.core.env.StandardEnvironment).
log4j:WARN Please initialize the log4j system properly.
log4j:WARN See http://logging.apache.org/log4j/1.2/faq.html#noconfig for more info.
instance me
init
true
destroydestroy
```

在 scope 为 singleton 时,每次使用 getBean 得到的都是同一个 bean,同一个对象,修改 bean 中的 scope 属性为"prototype"

控制台上显示的结果为:

```
log4j:WARN No appenders could be found for logger (org.springframework.core.env.StandardEnvironment).
log4j:WARN Please initialize the log4j system properly.
log4j:WARN See http://logging.apache.org/log4j/1.2/faq.html#noconfig for more info.
instance me
init
instance me
init
false
```

在 scope 为 prototype 时,每次得到的 bean 对象都是不同的。从上面可以看出实例化了两个对象,最终的比较是 false。

到这里,大家应该对 Spring 有了一个大致的印象,但只有了解其依赖注入实现的时候,才可以说开始了解 Spring 了。只有对其 AOP 实现有一定认识的时候,才能说,自己学会 Spring 了。首先,我们学习"依赖注入"。

8.4 依赖注入

8.4.1 什么是"依赖注入"

读一读下面这段有意思的对话：

路人甲：好多事情本来很简单,起了个硕大的名字后反而让人不知道是什么了。

路人乙：不起名字的话,你怎么知道我说的是哪个？就"那个、那个"地叫啊？

路人丙：名字是得起,别被那么大的名字唬住就成。

我们首先考虑：什么是依赖？两个元素中一个定义发生改变则会引起另一个元素发生变化,则称这两个元素之间存在依赖关系。一个类要发送消息给另一个类；一个类将另一个类作为其数据的一部分；一个类的操作中将另一个类作为其参数,这个类就依赖另一个类。

类与类之间的依赖关系增加了程序开发的复杂程度。我们在开发一个类的时候,还要考虑相关联的其他类的属性和行为。最开始的时候,程序只用来处理计算问题,没有复杂的逻辑,面向过程编程就可以搞定。后来,程序的规模和需要处理的逻辑越来越复杂,面向对象编程就有了必要,如图 8-6 所示。

图 8-6　面向对象使系统的实现变得容易

当我们用程序模拟一个柜子的组装过程时,面向对象的优势就明显地体现出来。软件需求的增长是没有止境的。当我们实现的系统越来越复杂,面向对象也显得苍白的时候,我们需要更多有效的理论和方法。如图 8-7 所示。

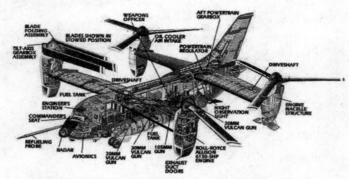

图 8-7 系统复杂程度越来越高,仅面向对象已经不够了

系统变得复杂是因为系统的各部分关联程度太高,即各模块间"依赖"程度太高。如果我们采取组件化的思想降低系统各组件的依赖关系,实现每个组件的时候只需要考虑这个组件需要实现的功能还有这个组件和其他部分的接口就可以了。

电脑是怎么制造出来的?有专门生产主板的厂商,有专门生产 CPU 的厂商,有专门生产硬盘的厂商……硬盘厂商设计产品的时候他需要考虑 CPU 要支持什么指令集,主板采用什么芯片组吗?当然不需要,他只知道别人会留好接口,他也会按照行业标准给别人留好接口。电脑的各个组件间当然是有依赖关系的,但由于明确地定义了接口,在各个组件实现的时候,完全不用考虑这些依赖关系,从而使我们获得了构建更复杂系统的能力。组件间的依赖关系和接口的重要性在将各个组件组装在一起的时候得以体现。通过这种开发模式,我们还获得了一种能力:我们可以随意地更换接口的实现,例如:采用什么显示器随你的便,实现了 VGA 接口即可,接个投影仪当家庭影院使都成!软件设计与此类似,Spring 提倡面向接口编程也是基于这样的考虑。所谓依赖注入,就是明确地定义组件接口(如 UserDAO),独立开发各个组件,然后根据组件间的依赖关系组装(UserAction 依赖于 UserBiz,UserBiz 依赖于 UserDAO)运行的设计开发模式。

系统组装的时候,只要是 UserDAO 的实现类就可以注入 UserBiz 中,可以是 UserDAO-JdbcImpl,也可以是 UserDAOHibimpl。究竟使用哪个实现类,组装的时候再决定。是不是很方便呢?

8.4.2 如何使用 Spring"依赖注入"

假设我们要开发一个打印机模拟程序:
1. 可以灵活地配置使用彩色墨盒还是灰色墨盒。
2. 可以灵活地配置打印机页面的大小。

程序中包括打印机(Printer)、墨盒(Ink)和纸张(Paper)3 个组件。打印机依赖墨盒和纸张。我们采取如下的步骤开发这个程序。

(1)定义 Ink 和 Paper 的接口。

(2)使用 Ink 和 Paper 接口开发 Printer 程序。在开发 Printer 时并不依赖 Ink 和 Paper 的实现类。

(3)开发 Ink 和 Paper 的实现类:ColorInk、GreyInk 和 TextPaper。

(4) 组装打印机，运行调试。
- 定义 Ink 和 Paper 接口，如示例代码 8-1 所示。

示例代码 8-1：Ink 接口源代码

```java
/*
 * 彩色墨盒
 */
public interface Ink {
    public String getColor(int r, int g, int b);
}
```

Ink 接口只定义一个 getColor 接口方法，传入红绿蓝三原色的值，表示逻辑颜色，返回一个形如 #ffc800 的颜色字符串，表示打印机采用的颜色。如示例代码 8-2 所示。

示例代码 8-2：Paper 接口源代码

```java
/*
 * 纸张接口
 */
interface Paper {
    public static final String newline = "\r\n";

    /*
     * 输出字符到纸张
     */
    public void putInChar(char c);

    /*
     * 得到输出到纸张上的内容
     */
    public String getContent();
}
```

Paper 接口中定义了两个方法，putInChar 用于向纸张中输出一个字符，向纸张输入字符后，纸张会根据自身大小（每页行数和每行字数的限制）在输入流中插入换行符、分页符和页码，getContent 用于得到纸张中的现有内容。
- 使用 Ink 接口和 Paper 接口开发 Printer，如示例代码 8-3 所示。

示例代码 8-3：Printer 类源代码

```java
/*
 * 打印机程序
 */
class Printer {
    public Ink ink = null;
    public Paper paper = null;

    public Ink getInk() {
            return ink;
    }

    public Paper getPaper() {
            return paper;
    }
    public void print(String str) {
            // 输出颜色标记
            System.out.println(" 使用 " + ink.getColor(255, 200, 0) + " 颜色打印 :\n");
            // 逐字符输出到纸张
            for (int i = 0; i < str.length(); ++i) {
                    paper.putInChar(str.charAt(i));
            }
            // 将纸张的内容输出
            System.out.print(paper.getContent());
    }
}
```

Printer 类中只有一个 print() 方法，输入参数是一个即将被打印的字符串，打印机将这个字符串逐个字符输入到纸张，然后将纸张中的内容输出。在开发 Printer 程序的时候，我们只需要了解 Ink 和 Paper 接口即可，完全不依赖这些接口的实现。设计硬盘的时候，设计师也只需要了解电源和数据接口的规范，不依赖具体的电源和主板的实现。在设计真实的打印机的时候也是这样的，设计师也是针对纸张和墨盒的接口规范进行设计。现在看示例代码 8-3 的程序。print() 方法运行的时候去哪里获得 Ink 和 Paper 的实例呢？

我们还需要提供"插槽"，以便组装的时候可以将 Ink 和 Paper 的实例"注入"进来，对 Java 代码来说就是 setter 方法。在示例代码 8-3 的 Printer 类代码中加入如示例代码 8-4 的两个 setter() 方法，Printer 类的开发工作就完成了。

示例代码 8-4：setter 方法源代码

```java
public void setInk(Ink ink) {
        this.ink = ink;
}

public void setPaper(Paper paper) {
        this.paper = paper;
}
```

- 开发 Ink 和 Paper 的实现类：ColorInk、GreyInk 和 TexPaper，如示例代码 8-5 所示。

示例代码 8-5：ColorInk.java 源代码

```java
package com.xtgj.j2ee.chapter08.spring;

import java.awt.Color;

/*
 * 彩色墨盒
 *
 * @author 迅腾国际
 */
public class ColorInk implements Ink {
    /*
     * (非 Javadoc)
     *
     * @see com.aptech.jb.ink.Ink#getColor(int, int, int)
     */
    public String getColor(int r, int g, int b) {
            Color color = new Color(r, g, b);
            return "#" + Integer.toHexString(color.getRGB()).substring(2);
    }
}
```

ColorInk 实现了 Ink 接口，覆盖了 getColor 方法，返回颜色值。如示例代码 8-6 所示。

示例代码 8-6：GreyInk.java 源代码

```java
/*
 * 灰色墨盒
 *
 * @author 迅腾国际
 */
class GreyInk implements Ink {
    public String getColor(int r, int g, int b) {
            int c = (r + g + b) / 3;
            Color color = new Color(c, c, c);
            return "#" + Integer.toHexString(color.getRGB()).substring(2);
    }
}
```

GreyInk 实现了 Ink 接口，覆盖了 getColor() 方法，返回颜色值。可以看到，彩色墨盒的 getColor() 方法只是把传入的颜色参数做了简单的格式转换；灰色墨盒则先把传入的颜色值进行计算，转换成灰度颜色，再进行格式转换。如示例代码 8-7 所示。

示例代码 8-7：TextPaper.java 源代码

```java
/*
 * 文本打印纸张实现
 *
 */
class TextPaper implements Paper {
    private String content = ""; // 纸张中内容
    private int charPerLine = 16;// 每行字符数
    private int linePerPage = 5;// 每页行数

    private int posX = 0;// 当前横向位置，从 0 到 charPerLine-1
    private int posY = 0;// 当前行数，从 0 到 linePerPage-1
    private int posP = 1;// 当前页数

    /*
     * ( 非 Javadoc)
     *
     * @see com.aptech.jb.Paper#getContent()
     */
    public String getContent() {
```

```java
            String ret = this.content;
            // 补齐本页空行，并显示页码
            if (!(posX == 0 && posY == 0)) {
                int count = linePerPage - posY;
                for (int i = 0; i < count; ++i) {
                    ret += Paper.newline;
                }
                ret += "== 第 " + posP + " 页 ==";
            }
            return ret;
        }


        /*
         * (非 Javadoc)
         *
         * @see com.aptech.jb.Paper#putInChar(char)
         */
        public void putInChar(char c) {
            content += c;
            ++posX;
            // 判断是否换行
            if (posX == charPerLine) {
                content += Paper.newline;
                posX = 0;
                ++posY;
            }
            // 判断是否翻页
            if (posY == linePerPage) {
                content += "== 第 " + posP + " 页 ==";
                content += Paper.newline + Paper.newline;
                posY = 0;
                ++posP;
            }
        }

        // Setter 方法，用于属性注入。
        public void setCharPerLine(int charPerLine) {
            this.charPerLine = charPerLine;
```

```
        }

    public void setLinePerPage(int linePerPage) {
        this.linePerPage = linePerPage;
    }

}
```

在 TextPaper 实现类的代码中,我们不用关心具体逻辑的实现。其中 content 用于保存当前纸张的内容。charPerLine 和 linePerPage 用于限定每行可以打印多少个字符和每页可以打印多少行。我们注意到 setCharPerLine 和 setLinePerPage 两个 setter 方法,跟示例代码 8-4 中的 setter 方法相同,这也是为了组装时"注入"数据留下的"插槽"。所以,我们不仅可以注入某个类的实例,还可以注入数值、字符串等基本数据类型。

● 组装打印机,运行调试

零件都准备好了,要开始组装啦!在前 3 步,我们都没有接触到 Spring,只有从组装这一步,才开始设计 Spring。首先,给项目添加 Spring 支持。右击项目名称节点,在弹出的右键菜单中选择"MyEclipse"→"Add Spring Capabilities…"选项:或者选中项目节点,从菜单里选择"MyEclipse"→"Add Spring Capabilities..."选项,如图 8-8 所示。在弹出的对话框中直接单击"下一步"或"完成"按钮。这个过程类似于前面的章节中添加 Hibernate 支持的方式,不赘述。

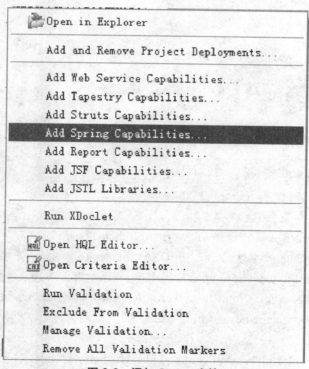

图 8-8 添加 Spring 支持

添加 Spring 支持成功后，MyEclipse 自动帮我们导入了 Spring 3.1 CoreLibraries 库，类库支持"依赖注入"，还自动在 src 目录下创建了 applicationContext.xml。applicationContext.xml 的内容如示例代码 8-8 所示。

示例代码 8-8：applicationContext.xml 源代码

```xml
<?xml version="1.0" encoding="UTF-8"?>
<beans
    xmlns="http://www.springframework.org/schema/beans"
    xmlns:xsi="http://www.w3.org/2001/XMLSchema-instance"
    xmlns:p="http://www.springframework.org/schema/p"
    xsi:schemaLocation="http://www.springframework.org/schema/beans   http://www.springframework.org/schema/beans/spring-beans-3.1.xsd">
</beans>
```

首先，创建几个待组装零件的实例，如图 8-9 所示。

图 8-9　待组装零件

示例代码 8-9：配置墨盒和纸张组件的源代码

```xml
<?xml version="1.0" encoding="UTF-8"?>
<beans
    xmlns="http://www.springframework.org/schema/beans"
    xmlns:xsi="http://www.w3.org/2001/XMLSchema-instance"
    xmlns:p="http://www.springframework.org/schema/p"
    xsi:schemaLocation="http://www.springframework.org/schema/beans   http://www.springframework.org/schema/beans/spring-beans-3.1.xsd">

    <bean id="colorInk" class="com.xtgj.j2ee.chapter08.spring.ColorInk" />
    <bean id="greyInk" class="com.xtgj.j2ee.chapter08.spring.GreyInk" />
```

```xml
<bean id="a4Paper" class="com.xtgj.j2ee.chapter08.spring.TextPaper">
    <property name="charPerLine" value="10" />
    <property name="linePerPage" value="8" />
</bean>
<bean id="b5Paper" class="com.xtgj.j2ee.chapter08.spring.TextPaper">
    <property name="charPerLine" value="10" />
    <property name="linePerPage" value="5" />
</bean>
</beans>
```

由示例代码 8-9 可知，在 Spring 的配置文件中，使用 <bean> 节点来创建 bean 的实例。这个节点有两个常用属性，一个是 id，表示定义的 bean 实例的名称；一个是 class，表示定义的 bean 的类型。

```xml
<bean id="colorInk" class="ColorInk" />
<bean id="greyInk" class="GreyInk" />
```
就相当于：
```java
ColorInk colorInk=new ColorInk();
GreyInk greyInk= GreyInk();
```
<bean> 节点有可选的 <property> 子节点，用于注入 bean 属性，以 a4Paper 为例：
```xml
<bean id="a4Paper" class="TextPaper">
    <property name="charPerLine" value="10" />
    <property name="linePerPage" value="5" />
</bean>
```
相当于：
```java
TextPaper a4Paper=new TextPaper();
a4Paper.setcharPerLine(10);
a4Paper.setlinePerPage (8);
```

上面定义 a4Paper 的 xml 代码相当于创建了一个名称为 a4Paper 的变量，并设置了它的两个属性的值。这也是刚才我们要创建 setCharPerLine() 和 setLinePerPage() 两个方法的原因。<property> 有两个属性：一个是 name，对应 bean 的 setter 方法声明的属性；另一个是 value，表示要注入的值。使用 Spring，类的创建和管理大多由 Spring 完成。"零件"都定义好了，现在看看打印机是怎么组装出来的。如示例代码 8-10 所示。

示例代码 8-10：配置打印机组件的源代码
```xml
<bean id="printer" class="com.xtgj.j2ee.chapter08.spring.Printer">
    <property name="ink" ref="colorInk"/>
    <property name="paper" ref="b5Paper"/>
</bean>
```

示例代码 8-10 组装了一台彩色的、使用 B5 打印纸的打印机。需要注意的是：<property> 的 value 属性没有了，而是采用了 ref 属性。value 属性用于注入基本类型的值。ref 属性用于注入已经定义好的 Bean，比如刚刚定义好的 colorInk、greyInk、a4Papel 和 b5Paper。由于 Printer 的 setInk(Ink ink)方法要求传入参数是 Ink(接口)类型，所以任何实现了 Ink 接口的类都可以注入。

同样，我们可以使用如下的定义组装一台灰色、使用 A4 打印纸的打印机。

```xml
<bean id="printer2" class="com.xtgj.j2ee.chapter08.spring.Printer">
    <property name="ink" ref="greyInk"/>
    <property name="paper" ref="a4Paper"/>
</bean>
```

完整的 applicationContext.xml 代码如示例代码 8-11 所示。

示例代码 8-11：完整的 applicationContext.xml 源代码

```xml
<?xml version="1.0" encoding="UTF-8"?>
<beans
    xmlns="http://www.springframework.org/schema/beans"
    xmlns:xsi="http://www.w3.org/2001/XMLSchema-instance"
    xmlns:p="http://www.springframework.org/schema/p"
    xsi:schemaLocation="http://www.springframework.org/schema/beans   http://www.springframework.org/schema/beans/spring-beans-3.1.xsd">

    <bean id="colorInk" class="com.xtgj.j2ee.chapter08.spring.ColorInk" />
    <bean id="greyInk" class="com.xtgj.j2ee.chapter08.spring.GreyInk" />
    <bean id="a4Paper" class="com.xtgj.j2ee.chapter08.spring.TextPaper">
        <property name="charPerLine" value="10" />
        <property name="linePerPage" value="8" />
    </bean>
    <bean id="b5Paper" class="com.xtgj.j2ee.chapter08.spring.TextPaper">
        <property name="charPerLine" value="10" />
        <property name="linePerPage" value="5" />
    </bean>

    <bean id="printer" class="com.xtgj.j2ee.chapter08.spring.Printer">
        <property name="ink" ref="colorInk"/>
        <property name="paper" ref="b5Paper"/>
    </bean>

</beans>
```

我们已经组装好了打印机,那么,组装好的"打印机"如何运行呢?测试代码如示例代码 8-12 所示。

示例代码 8-12:运行打印机

```java
package com.xtgj.j2ee.chapter08.spring;

import org.springframework.context.ApplicationContext;
import org.springframework.context.support.ClassPathXmlApplicationContext;

public class DITest {
    public static void main(String[] args) {
        ApplicationContext context = new ClassPathXmlApplicationContext(
                "applicationContext.xml");

        Printer printer = (Printer) context.getBean("printer");

        String str = " 几位轻量级容器的作者曾骄傲地对我说:这些容器非常有用,"
                + " 因为它们实现了 " 控制反转 "。这样的说辞让我深感迷惑:控制反转是 "
                + " 框架所共有的特征,如果仅仅因为使用了控制反转就认为这些轻量级 "
                + " 容器与众不同,就好像在说 " 我的轿车是与众不同的,因为它有四个轮子 "。";
        printer.print(str);
    }
}
```

运行示例代码 8-12,得到如图 8-10 所示的结果。

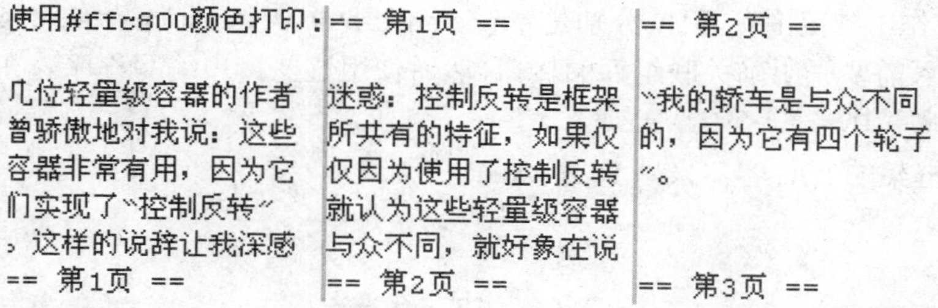

图 8-10 运行打印机

首先,Spring 读入 applicationContext.xml,根据 applicationContext.xml 的配置创建 Bean。

```
ApplicationContext context = new ClassPathXmlApplicationContext("applicationCon-
text.xml");
```

然后,通过 context.getBean(String beanName) 方法获得使用指定名称创建的 Bean。例如:按照示例代码 8-10 的配置,通过 context.getBean("b5Paper") 就可以得到 b5Paper 的实例引用。这里我们需要得到 printer 的引用。

```
Printer printer = (Printer) context.getBean("printer");
```

然后就可以直接调用 printer 的方法了。

通过示例演示,总结如下 Spring 有关系的只有组装(示例代码 8-11)和运行(示例代码 8-12)两部分。仅是这两部分代码就让我们获得了像更换打印机的墨盒和打印纸一样地更换程序组件的能力。这就是依赖注入带来的魔力!

通过 Spring 的强大组装能力,我们在开发每个程序组件的时候,只要明确关联组件的接口定义,不需要关心具体实现,这也就是所谓的"面向接口编程"。像前面的例子,得到彩色墨盒实例的"程序"代码如下:

```
Ink colorInk = (Ink)context.getBean("colorInk");
```

注意:组装的时候,class 属性的值是 ColorInk,但引用的时候,我们将永远使用 Ink 接口,坚持这样的"有效实践"会给我们的程序带来更大的灵活性。在前面的租房系统中,明确定义了 Biz 层和 DAO 层的接口,这也将给我们使用 Spring 重新组装程序的时候带来极大的便利。

这仅仅是开始,我们已经踏上了神奇的 Spring 之旅了!

8.5 小结

✓ Spring 是一个轻量级框架,提供依赖注入容器、AOP 实现、DAO/ORM 支持、Web 集成等功能。目标是使现有 Java EE 技术更易用,促进良好编程习惯。
✓ 三种实例化 bean 的方式分别是类构造器模式,静态工厂方式和实例工厂方式。
✓ Bean 的作用域范围有五种,分别是 singleton、prototype、session、request、global session。
✓ 依赖注入需要先明确关联组件的接口,然后使用这些接口编写程序。在运行前将接口的实现组装到程序中。
✓ Spring 框架的 IOC 容器采用 DI 和 AOP 实现。

8.6 英语角

inversion   反转

dependency	依赖
injection	注入
constructor	构造器
aspect	方面
oriented	面向
programming	编程

8.7 作业

定义接口 PersonDao 封装 save() 方法用于保存一个 Person 对象，定义该接口的实现类 PersonDaoBean，覆盖 save() 方法。定义服务类 PersonServiceBean，该类中定义 PersonDao 类型的对象 PersonDao，利用 Spring 的依赖注入技术实现控制反转，即应用本身不负责依赖对象 PersonDao 的创建及维护，依赖对象 PersonDao 的创建及维护是由外部容器负责的。这样控制权就由应用转移到了外部容器，控制权的转移就是所谓反转。

8.8 思考题

Spring 的核心技术包括什么？

8.9 学员回顾内容

1. Spring 的基本组成。
2. Spring 环境搭建。
3. 掌握实例化 bean 的三种方法。

第 9 章 Spring DI 详解

学习目标

- ✧ 了解 Spring DI 机制的基本原理
- ✧ 掌握 Spring 注入依赖对象的方式
- ✧ 了解 Spring 集合类型的装配
- ✧ 掌握注解方式注入依赖对象

课前准备

通过网络或其他途径了解实体和实体之间的关系,知道什么是依赖关系

本章简介

在上一章中,我们学习 Spring 框架的基本概念,知道了 Spring 框架的搭建方式,也简单了解了 Spring 框架的核心组成结构。它的"依赖注入"(Dependency Injection,DI)模式和"面向方面编程"(Aspect-Oriented programming,AOP)带给了编程者极大的方便。

在这一章中,我们将进一步讲解 Spring"依赖注入"的方式。被注入的对象都可以是什么样的数据类型?注入的形式又是怎样的?什么是手工装配方式和自动装配?这些问题都将在本章中得到解决。

9.1 Spring DI 机制的基本原理

9.1.1 Spring DI 原理举例

上一章中已经对"依赖注入"的应用做了简单介绍,我们再举个例子说明所谓的依赖。假设有一个类 Person,另一个类 Car,如果 Person 的开车方法 drive() 需要引用 Car,则称 Person 类依赖于 Car 类。延伸到对象,这种依赖关系依然成立,比如说 Person 类的对象 boy 依赖于 Car 类的对象 toyota。再讲讲这个 drive() 方法的实现,假定代码如下:

```java
public class Person{
// 省略部分代码
public void drive(){
Car toyota = new Car("TOYOTA");
toyota.挂挡；
toyota.踩油门；
toyota.打方向；
}
}
```

这其中的依赖关系，就导致了对象 boy 需要负责对象 toyota 的创建，甚至是整个生命周期的管理，而这样会造成代码耦合度高且不易维护等。比如说，要让这个男孩驾驶一辆 Audi，则还需要修改类 Person 的代码。因此在 Java 的设计理论中就提出了一条非常著名的原则，依赖倒转原则（Dependence Inversion），其核心思想就是要将这种具体类之间的依赖，尽量转换成抽象依赖。也就是说，类 Person 应该依赖于抽象类 ICar，而不是具体的类 Car。这里 Java 就大力推荐了抽象和接口的使用。

还是以上文的 boy 与 toyota 为例，其核心就是要将 boy 依赖的对象 toyota 注入到 boy 中去，而无需 boy 自己去引用 toyota。这个注入的过程，通常是由一个控制程序来完成的，无须对象去关心，举例如下：

```java
public class Person{
private ICar car;
public Person(ICar onecar){
car = onecar;
}
public void drive(){
car.挂挡；
car.踩油门；
car.打方向；
}
}
```

这个时候，进行注入并且调用的过程，就很简单了，如下：

```java
ICar toyota=new Toyota();
Person boy=new Person(toyota);
boy.drive();
```

注：这里我们假定，Toyota 类是 ICar 接口类的一个具体实现。

这个例子演示了一个最简单的注入方式的例子，也就是构造器方式注入，通过将依赖对象注入对象的构造器中来实现。当然，还有其他注入的方式，比如属性方式注入等。至此，我们对依赖注入的概念应该比较清楚了,那么 Spring 作为一个成熟的 IOC 容器如何实现 DI 的呢?

9.1.2 Spring 的 DI 机制

Spring 从核心而言,是一个 DI 容器,其设计哲学是提供一种无侵入式的高扩展性框架。即无需代码中涉及 Spring 专有类,却可将其纳入 Spring 容器进行管理。

为了实现无侵入性的目标,Spring 大量引入了 Java 的反射(Reflection)机制。通过动态调用的方式避免硬编码方式的约束,并在此基础上建立了其核心组件 BeanFactory,以此作为其依赖注入机制的实现基础。org.springframework.beans 包中包括了这些核心组件的实现类,核心中的核心为 BeanWrapper 和 BeanFactory 类。

- BeanWrapper 和动态设置属性

在运行期,由 Spring 根据配置文件,将其他对象的引用通过组件提供的 setter() 方法进行设定。利用反射(Reflection)机制实现动态设置对象属性,如下:

```
public static void diByReflection() throws Exception{
    Class cls = Class.forName("com.xtgj.j2ee.chapter09.demo. TheAction");
    Method mtd = cls.getMethod("setMessage",new Class[]{String.class});
    Object obj = (Object)cls.newInstance();
    mtd.invoke(obj,new Object[]{"Erica"});
}
```

利用 BeanWrapper 实现动态设置对象属性,原理一样,不过代码更简单了。如下:

```
public static void diByBeanWrapper() throws Exception {
    Object obj = Class.forName("com.xtgj.j2ee.chapter09.demo. TheAction").newInstance();
    BeanWrapper bw = new BeanWrapperImpl(obj);
    bw.setPropertyValue("message", "Erica");
    System.out.println("message=>"+bw.getPropertyValue("message"));
}
```

通过 BeanWrapper,我们无须在编码时就指定 JavaBean 的实现类和属性值,通过在配置文件加以设定,就可以在运行期动态创建对象并设定其属性(依赖关系)。

- BeanFactory 和配置文件

BeanWrapper 类实现了动态设置属性,但是未提及配置文件。解析配置文件的工作就由 BeanFactory 来完成。BeanFactory 代表 IOC 容器。BeanFactory 负责根据配置文件创建 Bean 实例,并对 Bean 进行管理,可以配置的项有:

(1)Bean 属性值及依赖关系(对其他 Bean 的引用)。
(2)Bean 创建模式(是否 Singleton 模式,即是否只针对指定类维持全局唯一的实例)。
(3)Bean 初始化和销毁方法。

- XmlBeanFactory

XmlBeanFactory 类是 BeanFactory 接口的一个最常用的实现。使用 XmlBeanFactory 实现如下：

```java
public static void diByBeanFactory() throws Exception {
    String fileName = " applicationContext.xml";
    Resource rs = new FileSystemResource(fileName);
    BeanFactory factory = new XmlBeanFactory(rs);
    Action action = (Action) factory.getBean("TheAction");
    System.out.println(action.execute("Rod Johnson"));
}
```

● bean 配置文件及说明如下：

```xml
<bean id="TheAction1" class=" com.xtgj.j2ee.chapter09.demo.UpperAction"
    init-method="init" destroy-method="cleanup"
    depends-on="ActionManager">
    <property name="message">
        <value>HeLLo</value>
    </property>
    <property name="desc">
    </property>
    <property name="dataSource">
        <ref local="dataSource" />
    </property>
</bean>
<bean id="dataSource"
    class="org.springframework.jndi.JndiObjectFactoryBean">
    <property name="jndiName">
        <value>java:comp/env/jdbc/sample</value>
    </property>
</bean>
```

◆ id：JavaBean 在 BeanFactory 中的唯一标识，代码中通过 BeanFactory 获取 JavaBean 实例时需以此作为索引名称。

◆ class：JavaBean 类名。

◆ init-method：初始化方法，此方法将在 BeanFactory 创建 JavaBean 实例之后，在向应用层返回引用之前执行。一般用于一些资源的初始化工作。

◆ destroy-method：销毁方法。此方法将在 BeanFactory 销毁的时候执行，一般用于资源释放。

◆ depends-on：Bean 依赖关系。一般情况下无须设定。Spring 会根据情况组织各个依赖关系的构建工作（这里示例中的 depends-on 属性非必须）。只有某些特殊情况下，如 JavaBean

中的某些静态变量需要进行初始化（这是一种 BadSmell，应该在设计上避免）。通过 depends-on 指定其依赖关系可保证在此 Bean 加载之前，首先对 depends-on 所指定的资源进行加载。

◆ <value>：通过 <value/> 节点可指定属性值。BeanFactory 将自动根据 JavaBean 对应的属性类型加以匹配。下面的"desc"属性提供了一个 null 值的设定示例。注意 <value></value> 代表一个空字符串，如果需要将属性值设定为 null，必须使用 <null/> 节点。

◆ <ref>：指定了属性对 BeanFactory 中其他 Bean 的引用关系。示例中，TheAction 的 dataSource 属性引用了 id 为 dataSource 的 Bean。BeanFactory 将在运行期创建 dataSource bean 实例，并将其引用传入 TheAction Bean 的 dataSource 属性。

● Resource

现在有 BeanWrapper 实现动态设置属性，有 BeanFactory 实现解析配置文件并创建 Bean 实例，DI 机制好像已经实现了。但是没有一个管理资源 Resource 的机制。

● ApplicationContext

我们发现 ApplicationContext 实现了类似 BeanFactory 的功能。ApplicationContext 在 BeanFactory 的基础上增加了其他功能，比如说 AOP 等。故推荐使用 ApplicationContext。

9.1.3 Spring 注入依赖对象的方式

Spring 注入基本类型对象时，可以采用构造器注入和属性 setter 方法注入两种方式，例如向 OrderServiceBean 中注入一个 String 类型的 name 属性，OrderServiceBean.java 源代码如示例代码 9-1 所示。

示例代码 9-1：OrderServiceBean.java 源代码

```
package com.xtgj.j2ee.chapter09.demo;

public class OrderServiceBean {
    private String name;

    public OrderServiceBean() {
    }

    public OrderServiceBean(String name) {
        this.name = name;
    }

    public String getName() {
        return name;
    }
```

```java
    public void setName(String name) {
        this.name = name;
    }

}
```

Spring 配置如示例代码 9-2 所示。

示例代码 9-2：applicationContext.xml 源代码

```xml
<?xml version="1.0" encoding="UTF-8"?>
<beans xmlns="http://www.springframework.org/schema/beans"
    xmlns:xsi="http://www.w3.org/2001/XMLSchema-instance"
    xsi:schemaLocation="http://www.springframework.org/schema/beans    http://www.springframework.org/schema/beans/spring-beans-2.5.xsd">
    <bean id="orderService" class="com.xtgj.j2ee.chapter09.demo.OrderServiceBean">
        <!-- 构造器注入 -->
        <constructor-arg index="0" type="java.lang.String" value="xxx" />
        <!-- 属性 setter 方法注入 -->
        <property name="name" value="zhao" />
    </bean>
</beans>
```

编写测试类 SpringTest.java，源代码如示例代码 9-3 所示。

示例代码 9-3：测试类 SpringTest.java 源代码

```java
package com.xtgj.j2ee.chapter09.test;

import org.junit.After;
import org.junit.Before;
import org.junit.Test;
import org.springframework.context.ApplicationContext;
import org.springframework.context.support.ClassPathXmlApplicationContext;

import com.xtgj.j2ee.chapter09.demo.OrderServiceBean;

public class SpringTest {

    ApplicationContext context = null;

    @Before
```

```java
        public void setUp() throws Exception {
            context = new ClassPathXmlApplicationContext("applicationContext.xml");
        }

        @After
        public void tearDown() throws Exception {
        }

        @Test
        public void testMethod1() {
            OrderServiceBean bean = (OrderServiceBean) context
                    .getBean("orderService");
            System.out.println(bean.getName());
        }
    }
```

Spring 注入其他 bean 对象时有两种方式，假设要向 OrderServiceBean 中注入一个 OrderDaoBean 的实例，OrderDaoBean.java 源代码如示例代码 9-4 所示。

示例代码 9-4：OrderDaoBean.java 源代码

```java
package com.xtgj.j2ee.chapter09.dao;

public class OrderDaoBean {
    public void save() {
        System.out.println("save order");
    }
}
```

方式一，修改 OrderServiceBean.java 源代码，添加一个 orderDao 属性，如下：

```java
public class OrderServiceBean {
    // 省略部分代码

    private OrderDaoBean orderDao;

    public OrderDaoBean getOrderDao() {
        return orderDao;
```

```
        }
        public void setOrderDao(OrderDaoBean orderDao) {
                this.orderDao = orderDao;
        }
}
```

在 Spring 配置文件中添加如示例代码 9-5 所示的配置。

示例代码 9-5：applicationContext.xml 源代码

```xml
<bean id="orderDao" class="com.xtgj.j2ee.chapter09.dao.OrderDaoBean" />
<bean id="orderService" class="com.xtgj.j2ee.chapter09.demo.OrderServiceBean">
    <property name="orderDao" ref="orderDao" />
</bean>
```

在测试类中添加测试方法 testMethod2()，代码如示例代码 9-6 所示。

示例代码 9-6：测试方法 testMethod2 源代码

```java
@Test
public void testMethod2() {
    OrderServiceBean bean = (OrderServiceBean) context.getBean("orderService");
    bean.getOrderDao().save();
}
```

方式二，内部 bean 方式，在 Spring 配置文件中添加如示例代码 9-7 所示的配置。

示例代码 9-7：applicationContext.xml 源代码

```xml
<bean id="orderService" class="com.xtgj.j2ee.chapter09.demo.OrderServiceBean">
    <property name="orderDao">
        <bean class="com.xtgj.j2ee.chapter09.dao.OrderDaoBean" />
    </property>
</bean>
```

注意：这个内部 bean 是不能够被其他的类使用的。测试方法同上。

9.1.4 Spring 集合类型的装配

（1）假设 OrderServiceBean 中包含如下的几个集合属性，代码如示例代码 9-8 所示。

示例代码 9-8：OrderServiceBea.java 源代码

```java
public class OrderServiceBean {
// 省略部分代码
```

```java
    private Set<String> sets = new HashSet<String>();
    private List<String> lists = new ArrayList<String>();
    private Properties properties = new Properties();
    private Map<String, String> maps = new HashMap<String, String>();

    public Set<String> getSets() {
        return sets;
    }

    public void setSets(Set<String> sets) {
        this.sets = sets;
    }

    public List<String> getLists() {
        return lists;
    }

    public void setLists(List<String> lists) {
        this.lists = lists;
    }

    public Properties getProperties() {
        return properties;
    }

    public void setProperties(Properties properties) {
        this.properties = properties;
    }

    public Map<String, String> getMaps() {
        return maps;
    }

    public void setMaps(Map<String, String> maps) {
        this.maps = maps;
    }
}
```

在 Spring 配置文件中添加如示例代码 9-9 所示的配置。

示例代码 9-9：applicationContext.xml 源代码

```xml
<bean id="orderService" class="com.xtgj.j2ee.chapter09.demo.OrderServiceBean">
    <property name="lists">
        <list>
            <value>Tom</value>
            <value>Lily</value>
            <value>Jacky</value>
        </list>
    </property>
    <property name="sets">
        <set>
            <value>Tom</value>
            <value>Lily</value>
            <value>Jacky</value>
        </set>
    </property>
    <property name="maps">
        <map>
            <entry key="Tom" value="28" />
            <entry key="Lily" value="29" />
            <entry key="Jacky" value="30" />
        </map>
    </property>
    <property name="properties">
        <props>
            <prop key="11">Tom</prop>
            <prop key="12">Lily</prop>
            <prop key="13">Jacky</prop>
        </props>
    </property>
</bean>
```

在测试类中添加测试方法 testMethod3，代码如示例代码 9-10 所示。

示例代码 9-10：测试方法 testMethod3 源代码

```java
@Test
public void testMethod3() {
    OrderServiceBean bean = (OrderServiceBean) context
            .getBean("orderService");
```

```java
        List<String> list = bean.getLists();
        for(String tmp : list){
            System.out.print(tmp+"\t");
        }
        System.out.println("\n-------------------------");

        Set<String> set = bean.getSets();
        for(String tmp : set){
            System.out.print(tmp+"\t");
        }
        System.out.println("\n-------------------------");

        Map<String, String> map = bean.getMaps();
        for(String key : map.keySet()){
            System.out.print(key+"\t"+map.get(key)+"\n");
        }
        System.out.println("\n-------------------------");

        Properties prop = bean.getProperties();
        for(Object key : prop.keySet()){
            System.out.print(key+"\t"+prop.get(key)+"\n");
        }
        System.out.println("\n-------------------------");
    }
```

运行测试代码,得到如图 9-1 的输出结果。

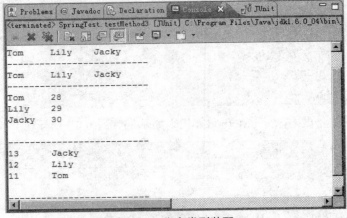

图 9-1　集合类型装配

9.2 注解方式注入依赖对象

利用 Spring 配置文件的配置方式注入依赖对象简单易懂，操作方便，那么还有没有更加简洁的方式能够实现这一功能呢？继 Spring 2.0 对 Spring MVC 进行重大升级后，Spring 2.5 又为 Spring MVC 引入了注解驱动功能。基于注解的配置有越来越流行的趋势，使用 Spring 的注解，我们无须让 Controller 继承任何接口，无需在 XML 配置文件中定义请求和 Controller 的映射关系，仅仅使用注解就可以让一个 POJO 具有 Controller 的绝大部分功能——Spring MVC 框架的易用性得到了进一步的增强。在框架灵活性、易用性和扩展性上，Spring MVC 已经全面超越了其他的 MVC 框架，伴随着 Spring 一路高唱猛进，可以预见 Spring MVC 在 MVC 市场上的吸引力将越来越不可抗拒。

采用注解方式注入依赖对象有两种方法，分别为手工装配和自动装配。这里，可以采用手工装配或自动装配两种方式注入依赖对象。当然，在实际应用中建议使用手工装配，因为自动装配会产生未知情况，开发人员无法预见最终的装配结果。

9.2.1 手工装配依赖对象

在 Java 代码中使用 @Autowired 或 @Resource 注解方式进行装配。但我们需要在 Spring 配置文件中配置如示例代码 9-11 所示的信息。

```xml
示例代码 9-11：applicationContext.xml 源代码
<?xml version="1.0" encoding="UTF-8"?>
<beans xmlns="http://www.springframework.org/schema/beans"
    xmlns:xsi="http://www.w3.org/2001/XMLSchema-instance"
    xmlns:context="http://www.springframework.org/schema/context"
    xsi:schemaLocation="
        http://www.springframework.org/schema/beans
        http://www.springframework.org/schema/beans/spring-beans-3.1.xsd
        http://www.springframework.org/schema/context
        http://www.springframework.org/schema/context/spring-context-3.1.xsd">

    <context:annotation-config />
</beans>
```

注解本身是无意义的，在 Spring 中需要配置四个注解后置处理，这个配置隐式注册了多个对注释进行解析处理的处理器：

- AutowiredAnnotationBeanPostProcessor
- CommonAnnotationBeanPostProcessor

✧ PersistenceAnnotationBeanPostProcessor
✧ RequiredAnnotationBeanPostProcessor

@Autowired 和 @Resource 这两个注解的区别是：@Autowired 默认按类型装配，@Resource 默认按名称装配，当找不到与名称匹配的 bean 时，才会按类型装配。

引入 @Autowired 注解（不推荐使用，建议使用 @Resource）类的实现（对成员变量进行标注）。例如，在上一节中，我们曾经向 OrderServiceBean 类中注入过 OrderDaoBean 的对象，用 @Autowired 注解便可以实现这样的效果。

注意，配置文件中的 orderService Bean 已经不需要再编写注入的代码了，"<context:annotationconfig />" 将隐式地向 Spring 容器注册注解处理器，这时，在 OrderServiceBean 类中添加 OrderDaoBean 对象的注解，如示例代码 9-12 所示。

示例代码 9-12：OrderServiceBean.java 源代码

```
public class OrderServiceBean {
    @Autowired       // 注解用于字段上
    private OrderDaoBean orderDao;
}

public class OrderServiceBean {
    @Autowired       // 注解用于属性的 setter 方法上
    public void setOrderDao(OrderDaoBean orderDao) {
        this.orderDao = orderDao;
    }
}
```

@Autowired 注解是按类型装配依赖对象，默认情况下它要求依赖对象必须存在，如果允许 null 值，可以设置它 required 属性为 false。如果我们想使用按名称装配，可以结合 @Qualifier 注解一起使用。示例如下：

```
public class OrderServiceBean {
    @Autowired  @Qualifier("orderDao")    // 按名称装配
    private OrderDaoBean orderDao;
}
```

@Autowired 可以对成员变量、方法和构造函数进行标注，来完成自动装配的工作。以上两种不同实现方式中，虽然 @AutoWired 的标注位置不同，但是它们都会在 Spring 初始化 OrderServiceBean 这个 bean 时，自动装配 orderDao 这个属性，区别是：第一种实现中，Spring 会直接将 orderDao 类型的唯一一个 bean 赋值给 orderDao 这个成员变量；第二种实现中，Spring 会调用 setOrderDao 方法来将 orderDao 类型的唯一一个 bean 装配到 orderDao 这个属性。

@Resource（JSR-250 标准注解，推荐使用它来代替 Spring 专有的 @Autowired 注解）。Spring 不但支持自己定义的 @Autowired 注解，还支持几个由 JSR-250 规范定义的注解，它们

分别是 @Resource、@PostConstruct 以及 @PreDestroy。@Resource 的作用相当于 @Autowired，只不过 @Autowired 按 byType 自动注入，而 @Resource 默认按 byName 自动注入罢了。@Resource 有两个属性是比较重要的，分别是 name 和 type，Spring 将 @Resource 注解的 name 属性解析为 bean 的名字，而 type 属性则解析为 bean 的类型。所以如果使用 name 属性，则使用 byName 的自动注入策略，而使用 type 属性时则使用 byType 自动注入策略。如果既不指定 name 也不指定 type 属性，这时将通过反射机制使用 byName 自动注入策略。@Resource 装配顺序如下：

（1）如果同时指定了 name 和 type，则从 Spring 上下文中找到唯一匹配的 bean 进行装配，找不到则抛出异常；

（2）如果指定了 name，则从上下文中查找名称（id）匹配的 bean 进行装配，找不到则抛出异常；

（3）如果指定了 type，则从上下文中找到类型匹配的唯一 bean 进行装配，找不到或者找到多个，都会抛出异常；

（4）如果既没有指定 name，又没有指定 type，则自动按照 byName 方式进行装配；如果没有匹配，则回退为一个原始类型（orderDao）进行匹配，如果匹配则自动装配。

@Resource 注解和 @Autowired 一样，也可以标注在字段或属性的 setter() 方法上，但它默认按名称装配。名称可以通过 @Resource 的 name 属性指定，如果没有指定 name 属性，当注解标注在字段上，即默认取字段的名称作为 bean 名称寻找依赖对象；当注解标注在属性的 setter 方法上，即默认取属性名作为 bean 名称寻找依赖对象。如示例代码 9-13 所示。

示例代码 9-13：OrderServiceBean.java 源代码
```
public class OrderServiceBean {
    @ Resource(name="orderDao ")          // 注解用于字段上
    private OrderDaoBean orderDao;
}
```

注意：如果没有指定 name 属性，并且按照默认的名称仍然找不到依赖对象时，@Resource 注解会回退到按类型装配。但一旦指定了 name 属性，就只能按名称装配了。

9.2.2 自动装配依赖对象

以上我们介绍了通过 @Autowired 或 @Resource 来实现在 Bean 中自动注入的功能。在一个稍大的项目中，通常会有上百个组件，如果这些组件采用 XML 的 bean 定义来配置，显然会增加配置文件的体积，查找及维护起来也不太方便。Spring 2.5 为我们引入了组件自动装配机制。对于自动装配，读者只需要了解，并不推荐使用。自动装配可以在 XML 配置文件中采用如下方式进行设置，例如：

```
<bean id="bean1" class="......" autowire="byType"/>
<bean id=" bean2" class="......" autowire="byName"/>
<bean id=" bean3" class="......" autowire="byType"/>
```

autowire 属性取值如下：
- byType：按类型装配，可以根据属性的类型，在容器中寻找跟该类型匹配的 bean。如果发现多个，那么将会抛出异常。如果没有找到，即属性值为 null。
- byName：按名称装配，可以根据属性的名称，在容器中寻找跟该属性名相同的 bean，如果没有找到，即属性值为 null。
- constructor：与 byType 的方式类似，不同之处在于它应用于构造器参数。如果在容器中没有找到与构造器参数类型一致的 bean，那么将会抛出异常。
- autodetect：通过 bean 类的自省机制（introspection）来决定是使用 constructor 还是 byType 方式进行自动装配。如果发现默认的构造器，那么将使用 byType 方式。
- Spring 2.5 还为我们引入了组件自动扫描机制，它可以在类路径底下寻找标注了 @Component、@Service、@Controller、@Repository 注解的类，并把这些类纳入进 Spring 容器中管理。它的作用和在 XML 文件中使用 bean 节点配置组件是一样的。下面我们将介绍如何注解 Bean，从而从 XML 配置文件中完全移除 Bean 定义的配置。

（1）@Component（不推荐使用）

只需要在对应的类上加上一个 @Component 注解，就将该类定义为一个 Bean 了，如示例代码 9-14 所示。

示例代码 9-14：OrderServiceBean.java 源代码

```
@Component
public class OrderServiceBean {
    @Autowired
    private OrderDaoBean orderDao;
}
```

使用 @Component 注解定义的 Bean，默认的名称 (id) 是小写开头的非限定类名。如这里定义的 Bean 名称就是 orderServiceBean。我们也可以指定 Bean 的名称，语法如下：

```
@Component("orderService ")
public class OrderServiceBean {
    @Autowired
    private OrderDaoBean orderDao;
}
```

@Component 是所有受 Spring 管理组件的通用形式，Spring 还提供了更加细化的注解形式：@Repository、@Service、@Controller，它们分别对应存储层 Bean，业务层 Bean，和控制层 Bean。目前版本中，这些注解与 @Component 的语义是一样的，完全通用，在 Spring 以后的版本中可能会给它们追加更多的语义。所以，我们推荐使用 @Repository、@Service、@Controller 来替代 @Component。@Service 用于标注业务层组件、@Controller 用于标注控制层组件（如 Struts 中的 action）、@Repository 用于标注数据访问组件，即 DAO 组件。而 @Component 泛指组件，当组件不好归类的时候，我们可以使用这个注解进行标注。

（2）@Repository

用于标注数据访问组件，即 DAO 组件，例如：

```
@Repository ("orderDao")
public class OrderDaoBean {
    public void save() {
            System.out.println("save order");
    }
}
```

（3）@Service

用于标注业务层组件，例如：

```
@Service ("orderService ")
public class OrderServiceBean {
    @Resource
    private OrderDaoBean orderDao;
}
```

（4）@Controller

用于标注控制层组件（如 Struts 中的 Action），例如：

```
@Controller ("order")
public class OrderAction {
    @Resource
    private OrderServiceBean orderService;
}
```

这里必须注意的是，使用上述注解的前提是必须让这些注解工作起来，具体的实现方式是在 Spring 配置文件中加入如示例代码 9-15 所示的配置信息。

示例代码 9-15：applicationContext.xml 源代码

```
<?xml version="1.0" encoding="UTF-8"?>
<beans xmlns="http://www.springframework.org/schema/beans"
    xmlns:xsi=http://www.w3.org/2001/XMLSchema-instance
xmlns:context="http://www.springframework.org/schema/context"
     xsi:schemaLocation="http://www.springframework.org/schema/beans
       http://www.springframework.org/schema/beans/spring-beans-3.1.xsd
       http://www.springframework.org/schema/context
       http://www.springframework.org/schema/context/spring-context-3.1.xsd">
```

```xml
<context:component-scan base-package="com.xtgj.j2ee.chapter09"/>

</beans>
```

这里，所有通过 <bean> 元素定义配置内容已经被移除，仅需要添加一行 <context:component-scan /> 配置就可以解决所有问题——Spring XML 配置文件得到了极致的简化（当然配置元数据还是需要的，只不过以注释形式存在罢了）。<context:component-scan /> 的 base-package 属性指定了需要扫描的类包，类包及其递归子包中所有的类都会被处理。<context:component-scan /> 还允许定义过滤器将基包下的某些类纳入或排除。Spring 支持以下 4 种类型的过滤方式：

（1）注解 org.example.SomeAnnotation 将所有使用 SomeAnnotation 注解的类过滤出来；
（2）类名指定 org.example.SomeClass 过滤指定的类；
（3）正则表达式 com\.kedacom\.spring\.annotation\.web\..* 通过正则表达式过滤一些类；
（4）AspectJ 表达式 org.example..*Service+ 通过 AspectJ 表达式过滤一些类。

以正则表达式为例：

```xml
<context:component-scan base-package="com.casheen.spring.annotation">
    <context:exclude-filter type="regex" expression="com\.casheen\.spring\.annotation\.web\..*" />
</context:component-scan>
```

值得注意的是 <context:component-scan /> 配置项不但启用了对类包进行扫描以实施注释驱动 Bean 定义的功能，而且还启用了注释驱动自动注入的功能（即隐式地在内部注册了 AutowiredAnnotation BeanPost Processor 和 CommonAnnotationBeanPostProcessor）。因此，当使用 <context:component-scan /> 后，就可以将 <context:annotation-config /> 移除了。

9.3 小结

Spring 大量引入了 Java 的反射（Reflection）机制，通过动态调用的方式避免硬编码方式的约束，并在此基础上建立了其核心组件 BeanFactory，以此作为其依赖注入机制的实现基础。

✓ Spring 注入基本类型对象时，可以采用构造器注入和属性 setter 方法注入两种方式。

✓ 基于注解的配置有越来越流行的趋势，使用 Spring 的注解，我们无须让 Controller 继承任何接口，无须在 XML 配置文件中定义请求和 Controller 的映射关系，仅仅使用注解就可以让一个 POJO 具有 Controller 的绝大部分功能。

✓ 在 Java 代码中使用 @Autowired 或 @Resource 注解方式进行装配。

✓ Spring 2.5 引入了组件自动扫描机制，它可以在类路径底下寻找标注了 @Component、@Service、@Controller、@Repository 注解的类，并把这些类纳入 Spring 容器中管理。

9.4 英语角

controller	控制器
autowired	自动装配
resource	资源
service	服务
repository	存储库
component	组件
PoJo(Plain Old Java Objects)	简单的 Java 对象

9.5 作业

定义接口 PersonDao 封装 save() 方法用于保存一个 Person 对象,定义该接口的实现类 PersonDaoBean,覆盖 save() 方法。定义接口 PersonService,封装 add() 方法用于添加 Person 对象,定义该接口的实现类 PersonServiceBean,该类中注入 PersonDao 类型的对象 PersonDao,利用 Spring 的注解方式实现。

9.6 思考题

Spring DI 技术的机制是什么?

9.7 学员回顾内容

1.Spring DI 机制的基本原理。
2.Spring 注入依赖对象的方式。
3.Spring 集合类型的装配。

第 10 章　Spring AOP

学习目标

- ◆ 了解 Spring AOP 机制的基本原理。
- ◆ 掌握 Spring AOP 的配置。
- ◆ 了解 Spring AOP 中的通知类型。
- ◆ 掌握 Spring AOP 的应用。

课前准备

通过网络或其他途径了解面向方面编程的特点,知道什么是方面、切点等。

10.1　面向方面编程概述

10.1.1　为什么需要 AOP

人脑的"内存"也是有限的,如果把所有相关联的事情像煮腊八粥一样一股脑放在一块儿,别说找出解决方案了,不会被逼疯的都是高人！通过依赖注入,在编写程序的时候,我们不用关心依赖的组件是怎么实现的。AOP 是从另外一个角度解决这一问题的,如图 10-1 所示。

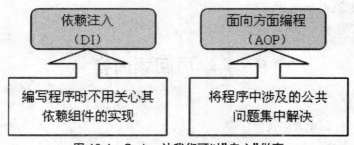

图 10-1　Spring 让我们可以"专心"做事

先来看下面的程序代码:

```java
public void doSomeBusiness(long lParam, String sParam) {
        // 记录日志
        log.info(" 调用 doSomeBusiness 方法,参数是：" + lParam);
        // 参数合法性验证
        if (lParam <= 0) {
                throw new IllegalArgumentException("xx 应该大于 0");

        }
        if (sParam == null || sParam.trim().equals("")) {
                throw new IllegalArgumentException("xx 不能为空 ");
        }
        // 异常处理
        try {
                // 真正的业务逻辑代码在这里 ...
                // 事务控制
                tx.commit();

        } catch (Exception e) {
                // ...
                tx.rollback();
        }
}
```

这是一个再"典型"不过的业务处理方法。日志、参数合法性验证、异常处理、事务控制等都是一个健壮的业务系统所必需的。否则,系统出现问题或有错误的业务操作时没日志可查。例如,传入的出库参数为负,出库反而导致库存增加;绩效计算到一半方法异常退出,已经插入的绩效记录没有回滚,重新启动计算则同一批货给代理商计算两次绩效……这样的系统显然是没人敢用的。甚至我们每时每刻都要关心这些方面的代码是否处理正确,权限控制是否正确,或者业务日志是否忘记做了,事务异常时是否加事务回滚的代码,出现 **SQLException** 时是否忘了在异常日志中记录 SQL 语句。

如果需要修改系统日志的格式,或者安全验证的策略,为了保证系统健壮可用,就要在每个业务方法里都反反复复地编写这些代码吗?考虑复杂的业务已经让人头大了,再处理这些方方面面的事情,写出来的代码既难读,质量又没保障。我们怎样才能专心在真正的业务逻辑上呢?这正是 AOP 要解决的问题!

10.1.2 什么是 AOP

AOP 是 Aspect-Oriented Progrmming 的缩写,意思是面向方面编程。在业务系统中,总有一些散落、渗透到系统各处而且不得不处理的事情,比如安全验证。我们需要在页面上判断用户是否登录、当前登录用户是否有权限访问该页面,在 Action 代码里还要限制用户是否通过直接在 URL 输入路径中绕过了页面的权限控制代码，甚至在业务层代码里，还要限制不同用户访问的不同数据。

与此类似，日志、事务、安全验证等这些通用的、散布在系统各处的需要在实现业务系统时关注的事情就称为"方面"，也称为"关注点"。如图 10-2 所示。如果能把这些"方面"集中处理，那样既能减少"方面"代码里的错误，又能保证我们在编写业务逻辑代码时专心做事，"方面"处理策略发生改变时还能统一做出修改，这就是 AOP 要做的事情。从系统中分离出方面，然后集中实现。从而可以独立编写业务逻辑代码和方面代码，在系统运行的时候，再将方面代码"织入"到系统中。就好比做衣服没有扣子和一些必需的饰品是不行的，那就让做衣服的去做衣服，生产扣子等饰品的去生产饰品。在衣服出厂前，再将这些饰品点缀到衣服上去。

同依赖注入一样，AOP 是一种设计思想，Spring 提供了一种优秀的实现。下面我们就来看一下 Spring 是怎么做到分离"方面"，并"织入"业务系统的。

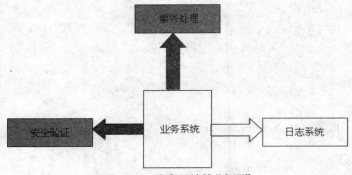

图 10-2　分离系统的"方面"

10.1.3　如何使用 AOP

假设我们要给一个购书系统的业务逻辑方法添加业务日志功能，要求在其业务方法调用前记录日志，记录方法调用的时间、调用的业务方法名和调用的参数。

需求分析：我们先给出业务方法的实现，这里的业务代码仅是单纯的业务逻辑，没有日志记录。因为在 AOP 的思想里，日志代码是单独实现的，不能加在业务方法代码中。我们可以在运行逻辑代码之前做"织入"工作，将"方面代码"织入到业务系统中。

● 实现模拟业务系统

仅模拟业务逻辑层的两个方法：buy(购书)和 comment(添加书评)。在使用 Spring 的时候，业务逻辑层也常被称作"服务层"。对应地，BookBiz 也可命名为 BookService。这仅是命名方式的不同，二者皆可，这里我们使用 Biz 的命名方式。

首先，定义在线图书销售系统业务逻辑接口，如示例代码 10-1 所示。

示例代码 10-1：在线图书销售系统业务逻辑接口

package com.xtgj.j2ee.chapter10.biz;

```java
/*
 * 在线图书销售系统业务逻辑接口
 *
 */
public interface BookBiz {
    public boolean buy(String userName, String bookName, double price);

    public void comment(String userName, String comments);
}
```

接着定义业务实现,如示例代码 10-2 所示。

示例代码 10-2:在线图书销售系统业务逻辑实现

```java
package com.xtgj.j2ee.chapter10.biz.impl;

import com.xtgj.j2ee.chapter10.biz.BookBiz;

public class BookBizImpl implements BookBiz {
    /*
     * 购买图书
     */
    public boolean buy(String userName, String bookName, double price) {
        System.out.println(" 业务方法 buy 开始执行 ");
        System.out.println("." + userName + " 购买图书:" + bookName);
        System.out.println("." + userName + " 增加积分:" + (int) (price / 10));
        System.out.println("." + " 向物流系统下发货单 ");
        System.out.println(" 业务方法 buy 结束 ");
        return true;
    }
    /*
     * 发表书评
     */
    public void comment(String userName, String comments) {
        System.out.println(" 业务方法 comment 开始执行 ");
        System.out.println("." + userName + " 发表书评 " + comments);
        System.out.println(" 业务方法 comment 结束 ");
    }
}
```

- 编写方面代码

实现特定功能的方面代码在 AOP 概念中又称为"通知"（Advice）。通知分为前置通知、后置通知、环绕通知、最终通知和异常通知。

这个分类是根据通知织入业务代码时执行的时间划分的。前置通知是在方法执行前自动执行的通知；后置通知是在方法执行后自动执行的通知；环绕通知能力最强，它可以在方法调用前执行通知代码，可以决定是否还调用目标方法；异常通知是方法抛出异常时自动执行的方面代码。具体图示如图 10-3 所示。

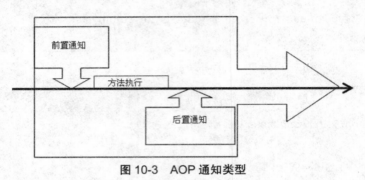

图 10-3　AOP 通知类型

我们这里使用前置通知，如示例代码 10-3 所示。

示例代码 10-3：前置通知代码

```java
package com.xtgj.j2ee.chapter10.log;

import java.lang.reflect.Method;
import java.text.DateFormat;
import java.text.SimpleDateFormat;
import java.util.Arrays;
import java.util.Date;

import org.springframework.aop.MethodBeforeAdvice;

public class LogAdvice implements MethodBeforeAdvice {
    private static DateFormat sdf = new SimpleDateFormat(
            "yyyy 年 mm 月 dd 日 hh 时 mm 分 ss 秒 ");

    public void before(Method m, Object[] args, Object target) throws Throwable {
        System.out.println("\n[ 系统日志 ][" + sdf.format(new Date()) + "]"
                + m.getName() + "(" + Arrays.toString(args) + ")");
    }
}
```

第 10 章 Spring AOP

编写前置通知需要实现 MethodBeforeAdvice 接口，这个接口要求实现的方法是：

> public void before(Method m,Object[]args,Object target) throw Throwable;

参数 m 是被通知目标方法对象，参数 args 是传入被调方法的参数，参数 target 是被调方法所属的对象实例。通过这些参数，我们几乎可以在方面代码中完成很多工作。

● 将方面代码织入业务对象中

如果直接访问原来的 Bean，通知代码肯定不会被执行。Spring 采用"代理"的方式将通知织入原 Bean 中，如图 10-4 所示。Spring 将原 Bean 和通知都封装到 org.springframework.aop.framework.proxyfactory 代理类中。用户通过访问代理类访问原 Bean，这样就能保证在目标方法调用前先执行前置通知的代码了。无须编写程序代码，只需要通过配置完成织入的过程即可，配置工作仍然是在 Spring 配置文件中完成的。通过代理方式织入如图 10-4。

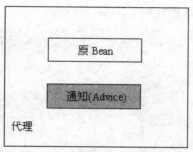

图 10-4 通过代理的方式将通知织入

在 applicationContext.xml 中添加如示例代码 10-4 所示的配置。

> 示例代码 10-4：AOP 的配置
>
> ```xml
> <?xml version="1.0" encoding="UTF-8"?>
> <beans
>     xmlns="http://www.springframework.org/schema/beans"
>     xmlns:xsi="http://www.w3.org/2001/XMLSchema-instance"
>     xmlns:p="http://www.springframework.org/schema/p"
>     xsi:schemaLocation="http://www.springframework.org/schema/beans      http://www.springframework.org/schema/beans/spring-beans-3.1.xsd">
>     <bean id="bookBizTarget"
>                 class="com.xtgj.j2ee.chapter10.biz.impl.BookBizImpl" />
>         <bean id="logAdvice" class="com.xtgj.j2ee.chapter10.log.LogAdvice" />
>         <bean id="bookBiz"
>             class="org.springframework.aop.framework.ProxyFactoryBean">
>             <property name="interceptorNames">
> ```

```xml
                        <list>
                            <value>logAdvice</value>
                        </list>
                    </property>
                    <property name="target" ref="bookBizTarget" />
            </bean>
    </beans>
```

首先定义了原 Bean "bookServiceTarget" 和通知 "logAdvice"。然后定义代理类，名称为 bookService，我们将通过这个 Bean 访问业务方法。代理类有 3 个必须设置的属性：proxyInterfaces 表示被代理的接口，interceptorNames 表示织入的通知列表，target 表示被代理的原 Bean。

这里设置属性值的方式和示例代码 10-4 略有不同。

```
表示属性的类型是 list 节点，其中 <value> 的节点可以有多个。
    <property name="InterfaceptorNames">
            <list>
                    <value>BookService</value>
            </list>
    </property>
    <property name="myGoodFirends">
            <list>
                    <value>Yang Bo</value>
                    <value>Xiao Lang</value>
                    <value>Dong li</value>
            </list>
    </property>
```

组装的工作完成了，运行如示例代码 10-5 所示的代码。

示例代码 10-5：测试类源代码

```java
package com.xtgj.j2ee.chapter10;

import org.springframework.context.ApplicationContext;
import org.springframework.context.support.ClassPathXmlApplicationContext;

import com.xtgj.j2ee.chapter10.biz.BookBiz;

public class AOPTest {
```

```
/**
 * @param args
 */
public static void main(String[] args) {
    ApplicationContext context = new ClassPathXmlApplicationContext(
            "applicationContext.xml");
    BookBiz bookBiz = (BookBiz) context.getBean("bookBizTarget");
    bookBiz.buy(" 高志水 ","CMMi 实务手册 ", 50);
    bookBiz.comment(" 王筝 ","《盗墓笔记》一点都不恐怖,很好看! ");
}
}
```

哪里有方面代码的痕迹啊？无论是业务方法代码中,还是调用业务方法的代码中,都看不到日志代码的蛛丝马迹。可运行的时候,确实输出了日志代码(如图 10-5 所示)。通过使用 Spring AOP,我们将日志代码分离出去,通过简单的配置,业务系统就具有了日志的能力！

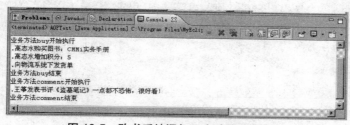

图 10-5　购书系统添加日志功能运行结果

10.2　AOP 详解

10.2.1　AOP 中的概念

● Aspect(方面):方面指横切性关注点的抽象,它与类相似,只是两者的关注点不一样。类是对物体特征的抽象,而方面是对横切性关注点的抽象。

● JoinPoint(连接点):所谓连接点是指那些被拦截到的点。在 Spring 中,这些点指的是方法,因为 Spring 只支持方法类型的连接点(实际上连接点还可以是 Field 或类构造器)。

● Pointcut(切入点):所谓切入点是指我们要对哪些连接点进行拦截的定义。

● Advice(通知):所谓通知是指拦截到连接点之后所要做的事情。通知分为前置通知,后置通知,异常通知,最终通知和环绕通知。

● Target(目标对象):代理的目标对象。

● Weave(织入):织入是指将切面应用到目标对象并导致代理对象创建的过程。

● Introduction（引入）：在不修改类代码的前提下，Introduction 可以在运行期为类动态地添加一些方法或 Field。

10.2.2 使用 Spring 进行面向方面编程

要进行 AOP 编程，首先我们要在 Spring 的配置文件中引入 aop 命名空间，如示例代码 10-6 所示。

示例代码 10-6：aop 命名空间

```xml
<?xml version="1.0" encoding="UTF-8"?>
<beans
    xmlns="http://www.springframework.org/schema/beans"
    xmlns:xsi="http://www.w3.org/2001/XMLSchema-instance"
    xmlns:p="http://www.springframework.org/schema/p"
    xmlns:aop="http://www.springframework.org/schema/aop"
    xsi:schemaLocation="http://www.springframework.org/schema/beans  http://www.springframework.org/schema/beans/spring-beans-3.1.xsd
        http://www.springframework.org/schema/aop  http://www.springframework.org/schema/aop/spring-aop-3.1.xsd">

</beans>
```

Spring 提供了两种方面声明方式，实际工作中我们可以选用其中一种：

● 基于注解方式声明方面

首先，启动对 @Aspect 注解的支持，在 Spring 配置文件中添加配置如示例代码 10-7 所示。

示例代码 10-7：启动对 @Aspect 注解的支持

```xml
<?xml version="1.0" encoding="UTF-8"?>
<beans
    xmlns="http://www.springframework.org/schema/beans"
    xmlns:xsi="http://www.w3.org/2001/XMLSchema-instance"
    xmlns:p="http://www.springframework.org/schema/p"
    xmlns:aop="http://www.springframework.org/schema/aop"
    xsi:schemaLocation="http://www.springframework.org/schema/beans  http://www.springframework.org/schema/beans/spring-beans-3.1.xsd
        http://www.springframework.org/schema/aop  http://www.springframework.org/schema/aop/spring-aop-3.1.xsd">

    <aop:aspectj-autoproxy proxy-target-class="true"/>
```

```xml
    <bean id="orderservice" class="com.xtgj.j2ee.chapter10.impl.OrderServiceBean"/>
    <bean id="log" class="com.xtgj.j2ee.chapter10.aspect.LogPrint"/>

</beans>
```

接着,基于注解方式声明方面,如示例代码10-8所示。

示例代码10-8:声明切面

```java
package com.xtgj.j2ee.chapter10.aspect;

import org.aspectj.lang.ProceedingJoinPoint;
import org.aspectj.lang.annotation.After;
import org.aspectj.lang.annotation.AfterReturning;
import org.aspectj.lang.annotation.AfterThrowing;
import org.aspectj.lang.annotation.Around;
import org.aspectj.lang.annotation.Aspect;
import org.aspectj.lang.annotation.Before;
import org.aspectj.lang.annotation.Pointcut;

@Aspect
public class LogPrint {
    @Pointcut("execution(* com.xtgj.j2ee.chapter10..*.*(..))")
    private void anyMethod() {
    }// 声明一个切入点

    @Before("anyMethod() && args(userName)")
    // 定义前置通知
    public void doAccessCheck(String userName) {
        System.out.println("doAccessCheck");
    }

    @AfterReturning(pointcut = "anyMethod()", returning = "revalue")
    // 定义后置通知
    public void doReturnCheck(String revalue) {
        System.out.println("doReturnCheck");
    }
```

```java
@AfterThrowing(pointcut = "anyMethod()", throwing = "ex")
// 定义例外通知
public void doExceptionAction(Exception ex) {
        System.out.println("doExceptionAction");
}

@After("anyMethod()")
// 定义最终通知
public void doReleaseAction() {
        System.out.println("doReleaseAction");
}

@Around("anyMethod()")
// 环绕通知
public Object doBasicProfiling(ProceedingJoinPoint pjp) throws Throwable {
        System.out.println("doBasicProfiling");
        return pjp.proceed();
}
```

最后，编写测试代码，如示例代码 10-9 所示。

示例代码 10-9：测试类 AOPTest.java 源代码

```java
package com.xtgj.j2ee.chapter10;

import org.springframework.context.ApplicationContext;
import org.springframework.context.support.ClassPathXmlApplicationContext;

public class AOPTest {

    public static void main(String[] args) {
            ApplicationContext context = new ClassPathXmlApplicationContext(
                        "applicationContext.xml");
            OrderServiceBean orderService = (OrderServiceBean) context
                        .getBean("orderservice");
            orderService.save();
    }
}
```

运行结果如图 10-6 所示：

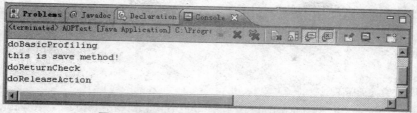

图 10-6　基于注解方式配置方面运行结果

使用 Spring，基本上不需要改动原来的代码，通过简单的配置就能达到目的。这也是 Spring 如此受大家喜爱的一个原因。在下一章，我们将学习应用范围更广的 AOP 应用：声明式事务。我们还将学习怎样使用 Spring 让 Hibernate 编码变得更简单，还有就是如何在 Web 应用中使用 Spring 组装我们之前编写的租房系统代码。完成这些工作后，SSH 集成框架的学习也将修成正果了。

- 基于 XML 配置方式声明方面

首先，在 Spring 的配置文件中引入 aop 命名空间。

接着，基于 XML 配置方式声明方面，方面代码如示例代码 10-10 所示。

```java
示例代码 10-10：LogPrint.java 源代码
package com.xtgj.j2ee.chapter10.aspect;

import org.aspectj.lang.ProceedingJoinPoint;

public class LogPrint {
    // 定义前置通知
    public void doAccessCheck(String userName) {
        System.out.println("doAccessCheck");
    }

    // 定义后置通知
    public void doReturnCheck(String revalue) {
        System.out.println("doReturnCheck");
    }

    // 定义例外通知
    public void doExceptionAction(Exception ex) {
        System.out.println("doExceptionAction");
    }
```

```java
// 定义最终通知
public void doReleaseAction() {
        System.out.println("doReleaseAction");
}

// 环绕通知
public Object doBasicProfiling(ProceedingJoinPoint pjp) throws Throwable {
        System.out.println("doReleaseAction");
        return pjp.proceed();
}
```

然后,定义接口 OrderService,源代码如示例代码 10-11 所示。

示例代码 10-11:接口 OrderService.java 源代码

```java
package com.xtgj.j2ee.chapter10;

/*
 * 接口
 *
 */
public interface OrderService {

    public void save();

}
```

再来定义该接口的实现类,源代码如示例代码 10-12 所示。

示例代码 10-12:实现类 OrderServiceBean.java 源代码

```java
package com.xtgj.j2ee.chapter10.impl;

import com.xtgj.j2ee.chapter10.OrderService;
/*
 * 实现类
 *
 */
```

```java
public class OrderServiceBean implements OrderService {

    public void save() {
            System.out.println(" this is save method! ");

    }
}
```

最后在 Spring 配置文件中添加如示例代码 10-13 所示代码的配置：

示例代码 10-13：applicationContext.xml 中的配置
```xml
<bean id="orderservice"
            class="com.xtgj.j2ee.chapter10.impl.OrderServiceBean" ></bean>
    <bean id="log" class="com.xtgj.j2ee.chapter10.aspect.LogPrint" />
    <aop:config>
            <aop:aspect id="myaop" ref="log">
                    <aop:pointcut id="mycut"
                            expression="execution(* com.xtgj.j2ee.chapter10..*.*(..))" />
                    <aop:before pointcut-ref="mycut" method="doAccessCheck" />
                    <aop:after-returning pointcut-ref="mycut"
                            method="doReturnCheck " />
                    <aop:after-throwing pointcut-ref="mycut"
                            method="doExceptionAction" />
                    <aop:after pointcut-ref="mycut" method="doReleaseAction" />
                    <aop:around pointcut-ref="mycut" method="doBasicProfiling" />
            </aop:aspect>
    </aop:config>
```

10.3 Spring+JDBC 组合开发

　　JdbcTemplate 是 core 包的核心类。它替我们完成了资源的创建以及释放工作，从而简化了 JDBC 的使用，通过传递一个 DataSource 引用来完成 JdbcTemplate 的实例化。
　　JdbcTemplate 的常用方法包括：
　　◇ update 方法：可以执行 insert、update 和 delete 语句。
　　◇ queryForXXX 方法：可以执行 select 语句。

10.3.1 使用 Spring+JDBC 集成步骤

使用 Spring+JDBC 集成步骤如下:
- 第一步,配置数据源,代码如示例代码 10-14 所示。

示例代码 10-14:配置数据源

```xml
<context:property-placeholder location="classpath:jdbc.properties"/>
<bean id="dataSource" class="org.apache.commons.dbcp.BasicDataSource" destroy-method="close">
    <property name="driverClassName" value="${driverClassName}"/>
    <property name="url" value="${url}"/>
    <property name="username" value="${username}"/>
    <property name="password" value="${password}"/>
    <!-- 连接池启动时的初始值 -->
    <property name="initialSize" value="${initialSize}"/>
    <!-- 连接池的最大值 -->
    <property name="maxActive" value="${maxActive}"/>
    <!-- 最大空闲值.当经过一个高峰时间后,连接池可以慢慢将已经用不到的连接慢
慢释放一部分,一直减少到 maxIdle 为止 -->
    <property name="maxIdle" value="${maxIdle}"/>
    <!-- 最小空闲值.当空闲的连接数少于阀值时,连接池就会预申请去一些连接,以
免洪峰来时来不及申请 -->
    <property name="minIdle" value="${minIdle}"/>
</bean>
```

- 第二步,配置事务。

配置事务时,需要在 XML 配置文件中引入用于声明事务的 tx 命名空间。事务的配置方式有两种:注解方式和基于 XML 配置方式。

在 Spring 配置文件中引入用于声明事务的 tx 命名空间,如示例代码 10-15 所示。

示例代码 10-15:配置事务引入命名空间

```xml
<beans xmlns="http://www.springframework.org/schema/beans"
    xmlns:xsi="http://www.w3.org/2001/XMLSchema-instance"
    xmlns:aop="http://www.springframework.org/schema/aop"
    xmlns:context="http://www.springframework.org/schema/context"
    xmlns:tx="http://www.springframework.org/schema/tx"
    xsi:schemaLocation="http://www.springframework.org/schema/beans
        http://www.springframework.org/schema/beans/spring-beans-3.1.xsd
```

```
            http://www.springframework.org/schema/tx http://www.springframework.org/
schema/tx/spring-tx-3.1.xsd
            http://www.springframework.org/schema/aop http://www.springframework.org/
schema/aop/spring-aop-3.1.xsd
            http://www.springframework.org/schema/context http://www.springframework.
org/schema/context/spring-context-3.1.xsd">
    ……
    </beans>
```

若采用注解方式配置事务,则在 Spring 配置文件中添加如下配置:

```
    <bean id="txManager" class="org.springframework.jdbc.datasource.DataSourceTransactionManager">
        <property name="dataSource" ref="dataSource"/>
    </bean>
    <!-- 采用 @Transactional 注解方式使用事务 -->
    <tx:annotation-driven transaction-manager="txManager"/>
```

相应地,需要在业务类中添加事务的注解,如下代码:

```java
@Service @Transactional
public class PersonServiceBean implements PersonService {
    private JdbcTemplate jdbcTemplate;
    @Resource
    public void setDataSource(DataSource dataSource) {
        this.jdbcTemplate = new JdbcTemplate(dataSource);
    }

    // 使用 JdbcTemplate 进行 insert/update/delete 操作
    public void save(Person person) throws Exception{
            jdbcTemplate.update("insert into person (name) values(?)",
            new Object[]{person.getName()}, new int[]{java.sql.Types.VARCHAR});
    }
    // 使用 JdbcTemplate 获取一条记录
    public Person getPerson(Integer id){
            RowMapper rowMapper = new RowMapper(){
                public Object mapRow(ResultSet rs, int rowNum) throws SQLException {
```

```java
                        Person person = new Person();
                        person.setId(rs.getInt("id"));
                        person.setName(rs.getString("name"));
                        return person;
                }
        };
        return (Person)jdbcTemplate.queryForObject("select * from person where id=?",
                        new Object[]{id}, new int[]{java.sql.Types.INTEGER}, rowMapper);
    }

    // 使用 JdbcTemplate 获取多条记录
    public List<Person> getPersons(){
            RowMapper rowMapper = new RowMapper(){
                    public Object mapRow(ResultSet rs, int rowNum) throws SQLException {
                            Person person = new Person();
                            person.setId(rs.getInt("id"));
                            person.setName(rs.getString("name"));
                            return person;
                    }
            };
            return jdbcTemplate.query("select * from person", rowMapper);
    }
}
```

若采用基于 XML 方式配置事务，则可以在 Spring 配置文件中添加如下配置：

```xml
<bean id="txManager" class="org.springframework.jdbc.datasource.DataSourceTransactionManager">
    <property name="dataSource" ref="dataSource"/>
</bean>
<aop:config>
    <aop:pointcut id="transactionPointcut" expression="execution(* com.xtgj.j2ee.chapter10.service..*.*(..))"/>
    <aop:advisor advice-ref="txAdvice" pointcut-ref="transactionPointcut"/>
</aop:config>
<tx:advice id="txAdvice" transaction-manager="txManager">
```

```xml
        <tx:attributes>
            <tx:method name="get*" read-only="true" propagation="NOT_SUPPORTED"/>
            <tx:method name="*"/>
        </tx:attributes>
    </tx:advice>
```

10.3.2 事务传播属性

● REQUIRED：业务方法需要在一个事务中运行。如果方法运行时，已经处在一个事务中，那么加入到该事务，否则为自己创建一个新的事务。

● NOT_SUPPORTED：声明方法不需要事务。如果方法没有关联到一个事务，容器不会为它开启事务。如果方法在一个事务中被调用，该事务会被挂起，在方法调用结束后，原先的事务便会恢复执行。

● REQUIRESNEW：属性表明不管是否存在事务，业务方法总会为自己发起一个新的事务。如果方法已经运行在一个事务中，则原有事务会被挂起，新的事务会被创建，直到方法执行结束，新事务才算结束，原先的事务才会恢复执行。

● MANDATORY：该属性指定业务方法只能在一个已经存在的事务中执行，业务方法不能发起自己的事务。如果业务方法在没有事务的环境下调用，容器就会抛出例外。

● SUPPORTS：这一事务属性表明，如果业务方法在某个事务范围内被调用，则方法成为该事务的一部分。如果业务方法在事务范围外被调用，则方法在没有事务的环境下执行。

● Never：指定业务方法绝对不能在事务范围内执行。如果业务方法在某个事务中执行，容器会抛出例外，只有业务方法没有关联到任何事务，才能正常执行。

● NESTED：如果一个活动的事务存在，则运行在一个嵌套的事务中。如果没有活动事务，则按 REQUIRED 属性执行。它使用了一个单独的事务，这个事务拥有多个可以回滚的保存点。内部事务的回滚不会对外部事务造成影响，它只对 DataSourceTransactionManager 事务管理器有效。

10.4 小结

✓ AOP 的目的是从系统中分离出方面，独立于业务逻辑实现，在程序执行时织入程序中运行。

✓ 方面代码在 AOP 中以通知的形式封装。

✓ 通知分前置通知、后置通知、环绕通知、异常通知和最终通知。

✓ Spring 配置文件是完成组装的主要场所，常用节点包括 <bean> 及其子节点 <property>。

✓ 横切性关注点的抽象即为方面，它与类相似，只是两者的关注点不一样，类是对物体特征的抽象，而方面横切性关注点的抽象。

- ✓ 连接点是指那些被拦截到的点。
- ✓ 切入点是指我们要对哪些连接点进行拦截的定义。
- ✓ 通知是指拦截到连接点之后所要做的事情。
- ✓ 织入是指将方面应用到目标对象并导致代理对象创建的过程。

10.5 英语角

aspect	方面
joinpoint	连接点
pointcut	切点
weave	织入
advice	通知
introduction	引入
target	目标对象

10.6 作业

定义接口 PersonService 封装 save() 方法用于保存一个 Person 对象,定义该接口的实现类 PersonServiceBean,覆盖 save() 方法。定义方面 CheckLogin,方面中声明各类通知,用于拦截未登录状态请求 save() 方法。可以采用注解方式和 XML 配置方式声明方面。

10.7 思考题

Spring AOP 技术的机制是什么？

10.8 学员回顾内容

1. Spring AOP 机制的基本原理。
2. Spring AOP 的配置。
3. Spring AOP 中的通知类型。

第 11 章　Spring 与 Struts、Hibernate 的集成

本章目标

- 掌握 Spring 与 Hibernate 的集成。
- 掌握 Spring 与 Struts 的集成。
- 掌握声明式事务。

课前准备

巩固 Struts、Hibernate、Spring 各个框架的单独使用

本章简介

前面我们学习了 Spring"依赖注入"和"面向方面编程"的原理和简单应用。这些技术怎样用到我们的项目中呢？本章我们就以租房系统中的用户模块为例，学习如何使用 Spring 简化 DAO 层代码、与 Hibernate 集成、管理依赖关系和管理事务。还要掌握 Spring 与 Struts 1 的集成，Spring 与 Struts 2 的集成将在上机部分讲述。

11.1　Spring 与 Hibernate 的集成

在前面章节里，我们介绍了使用 BaseHibernateDAO 简化 Hibernate 编码的方法，Spring 也提供了类似的支持，而且更方便、灵活，接口也更丰富。

现在，我们就试着使用 Spring 对 Hibernate 的支持重新实现租房信息管理系统中的用户模块。我们在前面章节示例中通过继承 BaseHibernateDAO 通用类，实现 UserDAO 这个接口。这次我们将通过继承 org.springframework.orm.hibernate3.HibernateDaoSupport 类，来实现这个接口。

UserDAO 接口的源代码如示例代码 11-1 所示。

示例代码 11-1：UserDAO 接口的源代码

```java
package com.xtgj.j2ee.zf.dao;

import java.util.List;

import com.xtgj.j2ee.zf.entity.TblUser;

/*
 * 用户 DAO 接口
 */
public interface UserDAO {
    /*
     * 根据条件查找用户
     * @param condition
     * @return
     */
    public List search(TblUser condition);
}
```

实现类 UserDAOHibimpl 源代码如示例代码 11-2 所示。

示例代码 11-2：UserDAOHibimpl 源代码

```java
package com.xtgj.j2ee.zf.dao.hibimpl;

import java.util.List;

import org.springframework.orm.hibernate3.support.HibernateDaoSupport;

import com.xtgj.j2ee.zf.dao.UserDAO;
import com.xtgj.j2ee.zf.entity.TblUser;

/*
 * 用户 DAO Hibernate 实现类
 *
 */
```

第 11 章 Spring 与 Struts、Hibernate 的集成

```java
public class UserDAOHibimpl extends HibernateDaoSupport implements UserDAO {

    public List search(TblUser condition) {
            List ret = super.getHibernateTemplate().findByExample(condition);
            return ret;
    }

}
```

在本章以后，大部分 DAO 的实现类都是这样实现的。在示例代码 11-2 中，我们没有看到创建 SessionFactory 的代码，这是因为 Spring 为这样处理的 HibernateDaoSupport 类提供了 setSessionFactory 方法。我们将通过 setter 方法向 DAO 类注入 SessionFactory。下面我们就看看配置数据源和 SessionFactory 具体是怎么实现的。

打开 Spring 的配置文件，首先要定义一个 SessionFactoryBean，才能将该 Bean 注入到 DAO 类中。如示例代码 11-3 所示，使用 Spring 提供的 LocalSessionFactoryBean，传入 Hibernate 配置文件的位置。

示例代码 11-3：在 Spring 配置文件中定义 SessionFactoryBean

```xml
<?xml version="1.0" encoding="UTF-8"?>
<!DOCTYPE beans PUBLIC "-//SPRING//DTD BEAN//EN" "http://www.springframework.org/dtd/spring-beans.dtd">

<beans>
    <!-- sessionFactory -->
    <bean id="sessionFactory"
            class="org.springframework.orm.hibernate3.LocalSessionFactoryBean">
        <property name="configLocation">
            <value>classpath:hibernate.cfg.xml</value>
        </property>
    </bean>

    <bean id="hib3TransactionManager"
            class="org.springframework.orm.hibernate3.HibernateTransactionManager">
        <property name="sessionFactory" ref="sessionFactory" />
    </bean>
    <!-- 声明式事务代理模板 -->
    <bean id="txProxyTemplate" abstract="true"
```

```xml
                    class="org.springframework.transaction.interceptor.TransactionProxyFactoryBean">
            <property name="transactionManager"
                    ref="hib3TransactionManager" />
            <property name="transactionAttributes">
                <props>
                    <prop key="add*">PROPAGATION_REQUIRED</prop>
                    <prop key="save*">PROPAGATION_REQUIRED</prop>
                    <prop key="delete*">PROPAGATION_REQUIRED</prop>
                    <prop key="update*">PROPAGATION_REQUIRED</prop>
                    <prop key="do*">PROPAGATION_REQUIRED</prop>
                    <prop key="*">PROPAGATION_REQUIRED,readOnly</prop>
                </props>
            </property>
        </bean>
        ……
    </beans>
```

在系统的开发阶段建议这样实现,当生成新的实体类的时候,会自动添加到 Hibernate 配置文件中,这时不需要手动修改 Spring 的配置文件。但这样做的缺点是无法有效地和 DataSource 进行集成。因此,在部署到"生产机"的时候,我们需要采用另外一种配置方法。如示例代码 11-4 所示。

示例代码 11-4: 在 Spring 配置文件中定义 SessionFactoryBean

```xml
    <!-- sessionFactory -->
    <bean id="dataSource" class="com.mchange.v2.c3p0.ComboPooledDataSource" destroy-method="close">
        <property name="driverClass" value="com.microsoft.sqlserver.jdbc.SQLServerDriver"/>
        <property name="jdbcUrl" value="jdbc:sqlserver://localhost:1433;DatabaseName=zf"/>
        <property name="user" value="sa"/>
        <!--<property name="password" value="123456"/>
        --><!-- 初始化时获取的连接数,取值应在 minPoolSize 与 maxPoolSize 之间。Default: 3 -->
        <property name="initialPoolSize" value="1"/>
        <!-- 连接池中保留的最小连接数。 -->
        <property name="minPoolSize" value="1"/>
        <!-- 连接池中保留的最大连接数。Default: 15 -->
        <property name="maxPoolSize" value="300"/>
```

```xml
        <!-- 最大空闲时间,60 秒内未使用则连接被丢弃。若为 0 则永不丢弃。Default: 0 -->
        <property name="maxIdleTime" value="60"/>
        <!-- 当连接池中的连接耗尽的时候 c3p0 一次同时获取的连接数。Default: 3 -->
        <property name="acquireIncrement" value="5"/>
        <!-- 每 60 秒检查所有连接池中的空闲连接。Default: 0 -->
        <property name="idleConnectionTestPeriod" value="60"/>
    </bean>

    <bean id="sessionFactory" class="org.springframework.orm.hibernate3.LocalSessionFactoryBean">
        <property name="dataSource" ref="dataSource"/>
        <property name="mappingResources">
            <list>
                <value>com/xtgj/j2ee/zf/entity/TblUser.hbm.xml</value>
            </list>
        </property>
        <property name="hibernateProperties">
            <value>
                hibernate.dialect=org.hibernate.dialect.SQLServerDialect
                hibernate.hbm2ddl.auto=create
                hibernate.show_sql=false
                hibernate.format_sql=false
            </value>
        </property>
    </bean>
```

已经定义好 SessionFactory，下面要将该实例注入 DAO 中。如示例代码 11-5 所示。

示例代码 11-5：SessionFactory 的注入

```xml
<!-- DAO -->
<bean id="userDAO"
        class="com.xtgj.j2ee.zf.dao.hibimpl.UserDAOHibimpl">
    <property name="sessionFactory" ref="sessionFactory" />
</bean>
```

11.2 使用 Spring 重新组装 Web 程序

11.2.1 使用 Spring 管理依赖关系

程序依赖关系是这样的：Action 依赖 Biz，Biz 依赖 DAO，DAO 依赖 SessionFactory，SessionFactory 依赖 DataSource。如果使用 Spring 管理这些依赖关系，注入的方向刚好与依赖的方向相反，如图 11-1 所示。

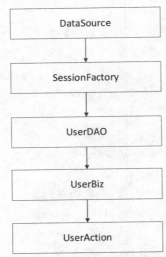

图 11-1　Java Web 应用中组件的注入关系

现在已经完成了 UserDAO 的定义。将 UserDAO 注入到 UserBiz 中需要这样做：

● 去掉直接实例化 UserDAO 接口实例的代码，这样就解除了 UserBiz 对 UserDAO 实现类的依赖关系，从而做到真正的"面向接口"编程。

● 增加 UserDAO 属性的 setter() 方法，方便注入。如示例代码 11-6 所示。

示例代码 11-6：UserBizImpl 源代码

```
package com.xtgj.j2ee.zf.biz.impl;

import java.util.List;

import com.xtgj.j2ee.zf.biz.UserBiz;
import com.xtgj.j2ee.zf.dao.UserDAO;
import com.xtgj.j2ee.zf.entity.TblUser;
```

```java
/*
 * 用户业务逻辑类实现
 *
 */
public class UserBizImpl implements UserBiz {
    private UserDAO userDAO = null;

    public boolean isExists(String userName) {
        // 返回值
        boolean bRet = false;
        // 封装查询条件
        TblUser condition = new TblUser();
        condition.setUname(userName);
        // 执行查询
        List ret = this.getUserDAO().search(condition);
        // 如果没有找到匹配的记录,返回 false
        if (ret == null || ret.size() == 0) {
            bRet = false;
        } else {
            bRet = true;
        }
        return bRet;
    }

    // Getters & Setters.
    public UserDAO getUserDAO() {
        return userDAO;
    }

    public void setUserDAO(UserDAO userDAO) {
        this.userDAO = userDAO;
    }
}
```

现在可以回到 Spring 配置文件中定义 UserBizBean。如示例代码 11-7 所示。

示例代码 11-7：在配置文件中定义 UserBizBean

```xml
<!-- BIZ -->
<bean id="userBizTarget"
    class="com.xtgj.j2ee.zf.biz.impl.UserBizImpl">
    <property name="userDAO" ref="userDAO" />
</bean>
<bean id="userBiz" parent="txProxyTemplate">
    <property name="target">
        <ref bean="userBizTarget" />
    </property>
</bean>
```

11.2.2 Spring 与 Struts 1 集成

我们知道，在 Struts 1 中 Action 类的实例是由 Struts 创建的。如果要使用 Spring 管理 Action 对 Biz 的依赖，那么 Action 类就要由 Spring 来创建。那么，怎么对 Struts 说"Spring 已经把 ActionBean 创建好了，拿过去用就可以了"呢？

Struts 可以以插件的方式扩展，利用 Spring 提供的 ContextLoaderPlugIn 插件即可做到 Spring 和 Struts 无缝集成。具体的步骤如下：

● 在 Struts 配置文件中配置插件，在 struts-config.xml 中加上如示例代码 11-8 所示的代码。

示例代码 11-8：Struts 配置文件中插件的配置

```xml
<?xml version="1.0" encoding="UTF-8"?>
<!DOCTYPE struts-config PUBLIC "-//Apache Software Foundation//DTD Struts Configuration 1.2//EN" "http://struts.apache.org/dtds/struts-config_1_2.dtd">

<struts-config>
    <form-beans>
        ……
    </form-beans>
    <action-mappings>
        ……
    </action-mappings>

    <message-resources
        parameter="com.xtgj.j2ee.zf.ApplicationResources" />
    <plug-in
```

第 11 章　Spring 与 Struts、Hibernate 的集成

```
                className="org.springframework.web.struts.ContextLoaderPlugIn">
                <set-property property="contextConfigLocation"
                        value="/WEB-INF/classes/applicationContext.xml" />
        </plug-in>
</struts-config>
```

在 struts-config.xml 中，需要把 Spring 配置文件的位置告知插件。classpath:applicationContext.xml 表示到 classpath 下去找 applicationContext.xml 文件。如果 Spring 配置文件放在 /WEB-INF/ 下，可以这样配置：value="/WEB-INF/applicationContext.xml"。如果有多个 Spring 配置文件，可以用逗号隔开，同时配置多个路径，如：value="/WEB-INF/applicationContext.xml,/WEB-INF/action.xml"。

● 将 Struts 配置文件中所有 Action 类的 type 属性值都配置为 "org.spring franqework.web.struts. Deleg ating ActionProxy" 类型，如示例代码 11-9 所示。

示例代码 11-9：Struts 配置文件中 Action 的配置

```
<action-mappings>
        <action path="/user" name="userForm" parameter="operate"
                type="org.springframework.web.struts.DelegatingActionProxy">
                <forward name="reg" path="/WEB-INF/jsp/user/reg.jsp" />
        </action>
</action-mappings>
```

● 在 Spring 配置文件中配置 UserAction，如示例代码 11-10 所示。注意这里 <bean> 没有 id 属性了，而是 name 属性，而且 name 属性的值要与 Struts 配置中相应 Action Bean 的 path 属性一致。

示例代码 11-10：Struts 配置文件中 UserAction 的配置

```
<!-- Action -->
<bean name="/user" class="com.xtgj.j2ee.zf.web.action.UserAction"
    >
    <property name="userBiz">
        <ref bean="userBiz" />
    </property>
</bean>
```

这样，程序中的依赖关系都使用 Spring 管理起来了。大名鼎鼎的 SSH 集成框架雏形初现！完整的 Spring 配置文件如示例代码 11-11 所示。

示例代码 11-11：完整的 Spring 配置文件

```xml
<?xml version="1.0" encoding="UTF-8"?>
<!DOCTYPE beans PUBLIC "-//SPRING//DTD BEAN//EN" "http://www.springframework.org/dtd/spring-beans.dtd">

<beans>
    <!-- sessionFactory -->
    <bean id="dataSource"
            class="com.mchange.v2.c3p0.ComboPooledDataSource"
            destroy-method="close">
        <property name="driverClass"
                value="com.microsoft.sqlserver.jdbc.SQLServerDriver" />
        <property name="jdbcUrl"
                value="jdbc:sqlserver://localhost:1433;DatabaseName=zf" />
        <property name="user" value="sa" />
        <!--<property name="password" value="123456"/>
        --><!-- 初始化时获取的连接数，取值应在 minPoolSize 与 maxPoolSize 之间。Default: 3 -->
        <property name="initialPoolSize" value="1" />
        <!-- 连接池中保留的最小连接数。-->
        <property name="minPoolSize" value="1" />
        <!-- 连接池中保留的最大连接数。Default: 15 -->
        <property name="maxPoolSize" value="300" />
        <!-- 最大空闲时间,60 秒内未使用则连接被丢弃。若为 0 则永不丢弃。Default: 0 -->
        <property name="maxIdleTime" value="60" />
        <!-- 当连接池中的连接耗尽的时候 c3p0 一次同时获取的连接数。Default: 3 -->
        <property name="acquireIncrement" value="5" />
        <!-- 每 60 秒检查所有连接池中的空闲连接。Default: 0 -->
        <property name="idleConnectionTestPeriod" value="60" />
    </bean>

    <bean id="sessionFactory"
            class="org.springframework.orm.hibernate3.LocalSessionFactoryBean">
        <property name="dataSource" ref="dataSource" />
        <property name="mappingResources">
            <list>
```

```xml
                    <value>com/xtgj/j2ee/zf/entity/TblUser.hbm.xml</value>
                </list>
        </property>
        <property name="hibernateProperties">
            <value>
                hibernate.dialect=org.hibernate.dialect.SQLServerDialect
                hibernate.hbm2ddl.auto=create hibernate.show_sql=false
                hibernate.format_sql=false
            </value>
        </property>
</bean>

<!--<bean id="sessionFactory"
        class="org.springframework.orm.hibernate3.LocalSessionFactoryBean">
        <property name="configLocation">
        <value>classpath:hibernate.cfg.xml</value>
        </property>
    </bean>

-->
<bean id="hib3TransactionManager"
        class="org.springframework.orm.hibernate3.HibernateTransactionManager">
        <property name="sessionFactory" ref="sessionFactory" />
</bean>
<!-- 声明式事务代理模板 -->
<bean id="txProxyTemplate" abstract="true"
        class="org.springframework.transaction.interceptor.TransactionProxyFactoryBean">
        <property name="transactionManager"
            ref="hib3TransactionManager" />
        <property name="transactionAttributes">
            <props>
                <prop key="add*">PROPAGATION_REQUIRED</prop>
                <prop key="save*">PROPAGATION_REQUIRED</prop>
                <prop key="delete*">PROPAGATION_REQUIRED</prop>
                <prop key="update*">PROPAGATION_REQUIRED</prop>
                <prop key="do*">PROPAGATION_REQUIRED</prop>
```

```xml
                <prop key="*">PROPAGATION_REQUIRED,readOnly</prop>
            </props>
        </property>
</bean>

<!-- DAO -->
<bean id="userDAO"
        class="com.xtgj.j2ee.zf.dao.hibimpl.UserDAOHibimpl">
        <property name="sessionFactory" ref="sessionFactory" />
</bean>

<!-- BIZ -->
<bean id="userBizTarget"
        class="com.xtgj.j2ee.zf.biz.impl.UserBizImpl">
        <property name="userDAO" ref="userDAO" />
</bean>
<bean id="userBiz" parent="txProxyTemplate">
        <property name="target">
                <ref bean="userBizTarget" />
        </property>
</bean>

<!-- Action -->
<bean name="/user" class="com.xtgj.j2ee.zf.web.action.UserAction"
        >
        <property name="userBiz">
                <ref bean="userBiz" />
        </property>
</bean>
</beans>
```

在这个框架上开发新的表对应的功能时,在这个配置文件基础上依次增加 DAO、Biz、Service 和 Action 的配置即可。

11.3 声明式事务

11.3.1 应用系统中的事务管理

在前面几章的代码中，DAO 层进行事务控制。在企业应用系统中，这是不适当的。比如某图书馆的 IT 系统的业务逻辑层中，有这样一个业务方法：

```
public void lendBook(Student stu, Book book) {
    if (stu.getBorrowedBookCount() < 5) {
        stu.setBorrowBookCount(stu.getBorrowedBookCount() + 1);
        book.setStatus(" 已借出 ");
        book.setLender(stu);
        studentDAO.update(book);// 保存学生状态
        bookDAO.update(book);// 保存书的状态
        LendRecord lr = new LendRecord(stu, book);
        LendRecordDAO.add(lr);// 添加借阅记录
    } else {
        throw new Exception(" 每个学生只能借阅 5 本书 ");
    }
}
```

这个业务方法中执行了 3 个持久化操作，根据业务需求，要求这 3 个操作要么全部成功，要么全部失败。也就是，客观上需要对业务逻辑层的方法进行事务控制，而不是仅仅对 DAO 层的方法做事务控制。怎么实现呢？

事务处理是企业应用开发中不能回避的一个重要问题。在 Hibernate 风行之前，JDBC 是数据持久化的主要手段。为了达到在业务逻辑层进行事务控制，一个通用的做法是将 Connection（JDBC 中，事务依附在 Connection 对象上）对象以参数的方式传来传去。业务逻辑层应该只出现描述业务逻辑的代码，Connection 类在业务逻辑层的出现破坏了三层结构的基本原则。

另外一个做法是将业务逻辑放到 DAO 层实现，DAO 层依据业务逻辑创建方法，编写复杂的业务逻辑处理代码，这样进行架构设计的系统很快就会尝到苦果。当参加项目的人员达到一定数量，或者项目开发周期到一定时间，或者项目维护到一定时间的时候，人们发现这些系统的代码已经变得千头万绪了，剪不断，理还乱。在软件生产大工厂时代，质量是软件的生命线，是客户源源不断的重要保障，是大把大把人民币的重要前提，是免于周末加班，凌晨两点被电话吵醒，不得不去紧急解决客户问题的护身符，是看着别人瞎忙半个月也平不了一张报表，自己却可以悠闲地品口香茶、翻两页闲书时心头暗存的侥幸。一定要随时保证你的代码逻辑清晰、规范易读，不要抱着侥幸的心理破坏规矩。概念完整性说起来很玄虚，但当你违背它的时候，它的反面就会变得形象而具体。

在上一章我们学到，事务是系统开发过程中的一个"方面"，散布在系统里，但我们通过 AOP 技术可以将其放在一起实现。下面我们就学习在 Spring 中是如何通过声明的方式给系统增加事务支持的。

11.3.2 声明式事务

使用 Spring 的声明式事务，我们无须编写程序代码，所有的工作全在配置文件中完成。Spring 中提供的声明式事务的配置方法同上一章我们配置 logAdvice 的过程类似。首先，声明一个事务管理器，如示例代码 11-12 所示，需要注入 sessionFactory 属性。

示例代码 11-12：声明事务管理器

```
<bean id="txManager" class="org.springframework.orm.hibernate3.HibernateTransactionManager">
    <property name="sessionFactory" ref="sessionFactory"/>
</bean>
```

现在我们演示如何在前面 Spring 组装的基础上给 UserBiz 的方法加上事务支持。通过上一章 AOP 配置的方式我们知道，Spring AOP 的实现采用代理的方式，首先要定义被代理的对象，然后注入到代理类中。

首先，定义 UserBizTarget。然后，定义代理类 UserBiz，class 属性为 org.springframework.transaction.interceptor.TransactionProxyFactoryBean。将事务管理器（transactiomManager）和代理对象（target）注入。最后要做的就是对其事务属性（transactionAttributes）进行事务声明，如示例代码 11-13 所示。

示例代码 11-13：给 UserBiz 的方法加上事务支持

```xml
<bean id="userBizTarget" class="com.xtgj.j2ee.zf.biz.impl.UserBizImpl">
    <property name="userDAO" ref="userDAO" />
</bean>
<bean id="UserBiz"
    class="org.springframework.transaction.interceptor.TransactionProxyFactoryBean">
    <property name="transactionManager"
        ref="myHibTransactionManager" />
    <property name="target" ref="UserBizTarget" />
    <property name="transactionAttributes">
        <props>
            <prop key="add*">PROPAGATION_REQUIRED</prop>
            <prop key="del*">PROPAGATION_REQUIRED</prop>
```

```xml
        <prop key="update*">PROPAGATION_REQUIRED</prop>
        <prop key="do*">PROPAGATION_REQURED</prop>
        <prop key=" *">PROPAGATION_REQUIRED,readonly</prop>
      </props>
    </property>
</bean>
```

可以声明多条事务属性,每条事务属性是这样定义的:

```xml
<prop key="add*">PROPAGATION_REQUIRED</prop>
```

key 表示匹配方法的"模式",可以使用"*"通配符,例如,"add*"表示这条属性应用于 add 开头的方法。属性的值为 PROPAGATION_REQUIRED,表示当前方法必须运行在一个事务环境中。如果一个现有事务正在运行中,该方法将运行在这个事务中。否则,就要开始一个新的事务。可选的值如下:

（1）PROPAGATION_MANDATORY,表示当前方法必须运行在一个事务上下文中,否则就抛出异常。

（2）PROPAGATION_NEVER,表示当前方法不应该运行在一个事务上下文中,否则就抛出异常。

（3）PROPAGATION_REQUIRES_NEW,表示当前方法必须运行在自己的事务里。

（4）PROPAGATION_SUPPORTS,表示当前方法不需要事务处理环境,但如果有一个事务正在运行的话,这个方法也可以运行在这个事务里。

一般来讲,我们将业务逻辑层方法的事务属性设置为"PROPAGATION_REQUIRED"。但出于效率的考虑,对于需要更改数据库数据的方法也做如下配置:

```xml
<prop key="*">PROPAGATION'_REQUIRED,readonly</prop>
```

key="*"表示匹配所有其他的方法，readonly 表示启用只读事务。这样,数据库就可以采取合适的优化措施避免不必要的操作。

在示例代码 11-12 和示例代码 11-13 中,我们重新定义了 UserBiz,并为其声明了事务支持。其他代码不需要修改,我们可以照常将 UserBiz 注入需要它的 ActionBean 中。

想一想,系统中业务逻辑 Bean 的数量绝对不止一个,对每个业务逻辑 Bean 我们都要做这样的设置吗?那将是一个庞大的工作量。想想之前我们定义基类解决重复性编码的技巧,在 Spring 配置中也有类似的做法。首先,我们定义一个"基类",设置其 abstact 属性为 true,表示不可以直接得到其实例,如示例代码 11-14 所示。

示例代码 11-14：定义一个"基类"解决重复性编码

```xml
<!-- 声明式事务代理模板 -->
<bean id="txProxyTemplate" abstract="true"
        class="org.springframework.transaction.interceptor.TransactionProxyFactoryBean">
    <property name="transactionManager"
            ref="hib3TransactionManager" />
    <property name="transactionAttributes">
        <props>
            <prop key="add*">PROPAGATION_REQUIRED</prop>
            <prop key="save*">PROPAGATION_REQUIRED</prop>
            <prop key="delete*">PROPAGATION_REQUIRED</prop>
            <prop key="update*">PROPAGATION_REQUIRED</prop>
            <prop key="do*">PROPAGATION_REQUIRED</prop>
            <prop key="*">PROPAGATION_REQUIRED,readOnly</prop>
        </props>
    </property>
</bean>
```

然后，可以定义"继承"自这个基类的业务逻辑 Bean，很轻松地实现了多个业务逻辑 Bean 的定义，如示例代码 11-15 所示。

示例代码 11-15：定义"继承"自基类的业务逻辑 Bean

```xml
<!-- BIZ -->
<bean id="userBizTarget"
    class="com.xtgj.j2ee.zf.biz.impl.UserBizImpl">
    <property name="userDAO" ref="userDAO" />
</bean>
<bean id="userBiz" parent="txProxyTemplate">
    <property name="target">
        <ref bean="userBizTarget" />
    </property>
</bean>
```

这里，我们使用了 <bean> 的 parent 属性，通用的部分在 baseTransProxy 中配置，target 属性在"子 Bean"中配置。这就是 Spring 2.5 中分布式事务的配置方法。

11.3.3 Spring 2.5 声明事务的方式

采用 Spring 1.x 配置分布式事务要额外配置一个代理对象，原来的 Bean 的声明也要修改。因此，虽然没有侵入到程序代码中，仍然是一种"侵入式"的解决方案。

Spring 2.5 对分布式事务的配置方法进行了升级。配置过程更容易理解，而且可以在保持示例代码 11-7 不变的基础上给 UserBizBean 的方法增加事务支持。

首先，我们要升级项目中 Spring 的版本：

（1）添加 Spring 2.5 的 jar 包，包括 springjal、aspectjrt.jar、aspectjweaver.jar 和 cglib-nodep-2.1_3jar。

（2）修改 Spring 配置文件，增加常用命名空间的声明，如示例代码 11-16 所示。

示例代码 11-16：Spring 配置文件命名空间

```xml
<?xml version="1.0" encoding="UTF-8"?>
<beans xmlns="http://www.springframework.org/schema/beans"
    xmlns:xsi="http://www.w3.org/2001/XMLSchema-instance"
    xmlns:context="http://www.springframework.org/schema/context"
    xmlns:aop="http://www.springframework.org/schema/aop"
    xmlns:tx="http://www.springframework.org/schema/tx"
    xsi:schemaLocation="http://www.springframework.org/schema/beans
       http://www.springframework.org/schema/beans/spring-beans-2.5.xsd
       http://www.springframework.org/schema/context
       http://www.springframework.org/schema/context/spring-context-2.5.xsd
       http://www.springframework.org/schema/aop http://www.springframework.org/schema/aop/spring-aop-3.1.xsd
       http://www.springframework.org/schema/tx http://www.springframework.org/schema/tx/spring-tx-3.1.xsd">
```

重新启动项目，项目正常启动，表示项目中采用的 Spring 版本更新成功。如果是新项目采用 Spring 2.5 开发的话，就不需要使用 MyEclipse 工具给项目添加 Spring 支持了（当前 MyEclipse 版本中自带的 Spring 版本为 3.1），直接添加 Spring 3.1 的 jar 包和 3.1 风格的配置文件到项目中即可。

将示例代码 11-11 中配置文件按照示例代码 11-16 进行修改，然后我们开始 Spring 2.5 风格的声明式事务配置。无须改变原来 UserBiz 的配置，添加示例代码 11-17 中的代码到 Spring 配置文件中即可：

> **示例代码 11-17：Spring 配置文件修改**
>
> ```xml
> <tx:advice id="transactionAdvice" transaction-manager="transactionManager">
>     <tx:attributes>
>         <tx:method name="save*" propagation="REQUIRED"/>
>         <tx:method name="update*" propagation="REQUIRED"/>
>         <tx:method name="delete*" propagation="REQUIRED"/>
>         <tx:method name="remove*" propagation="REQUIRED"/>
>         <tx:method name="get*" propagation="NOT_SUPPORTED"/>
>         <tx:method name="validate*" propagation="NOT_SUPPORTED"/>
>         <tx:method name="find*" propagation="NOT_SUPPORTED"/>
>     </tx:attributes>
> </tx:advice>
> <aop:config>
>     <aop:pointcut id="transactionPointcut" expression="execution(*com.xtgj.j2ee.zf.biz..*.*(..))"/>
>     <aop:advisor advice-ref="transactionAdvice" pointcut-ref="transactionPointcut"/>
> </aop:config>
> ```

首先，通过 <tx:advice> 定义事务通知，需要指定一个事务管理器，然后在其属性中声明事务规则，含义与 1.x 风格配置中的类似。

然后，定义一个切点（pointcut，即定义哪些方法应用这些规则）：

> ```xml
> <aop:pointcut id="transactionPointcut" expression="execution(*com.xtgj.j2ee.zf.biz..*.*(..))"/>
> ```

上面的代码表示：com.xtgj.j2ee.zf.biz 包下的所有类的所有方法都应用事务规则。

最后，将事务通知和切点组合：

> ```xml
> <aop:advisor advice-ref="transactionAdvice" pointcut-ref="transactionPointcut"/>
> ```

是不是很简单呢？至此，我们已经获得一个成熟的三层架构+MVC 的框架了，Struts 主持大局，Hibernate 负责持久化工作，Spring 管理组件间的依赖并提供事务支持。

上述准备工作做完之后，我们可以很轻松地编写前端页面。本案例中的注册页面 reg.jsp 源代码如示例代码 11-18 所示。

示例代码11-18：注册页面 reg.jsp 源代码

```jsp
<%@ page language="java" contentType="text/html; charset=GBK" pageEncoding="GBK"%>

<%@ taglib uri="http://struts.apache.org/tags-bean" prefix="bean" %>
<%@ taglib uri="http://struts.apache.org/tags-html" prefix="html" %>
<%@ taglib uri="http://struts.apache.org/tags-logic" prefix="logic" %>

<HTML>
<HEAD>
<TITLE> 北京出租房 </TITLE>
<meta http-equiv="Content-Type" content="text/html; charset=GBK">

<link href="style/mycss.css" rel="stylesheet" type="text/css" />
<link href="style/texts.css" rel="stylesheet" type="text/css" />
<link href="style/btn.css" rel="stylesheet" type="text/css" />
<script lang="javascript">
    var xmlHttpRequest;
    function createXmlHttpRequest(){
            if(window.ActiveXObject){
                    return new ActiveXObject("Microsoft.XMLHTTP");
            }else if(window.XMLHttpRequest){
                    return new XMLHttpRequest();
            }
    }
    function checkUserExists(oCtl){
            var uname = oCtl.value;
            if (!uname){
                    alert(" 用户名不能为空 ");
                    oCtl.focus();
                    return ;
            }
            // 发送请求到服务器,判断用户名是否存在
            //var url = "?operate=doCheckUserExists&uname="+uname;
            var url = "?operate=toReg";
            xmlHttpRequest = createXmlHttpRequest();
```

```
            xmlHttpRequest.onreadystatechange = haoLeJiaoWo;
        xmlHttpRequest.open("GET",url,true);
        xmlHttpRequest.send(null);
    }
    function haoLeJiaoWo(){
            if(xmlHttpRequest.readyState == 4){
        if(xmlHttpRequest.status == 200){
            var b = xmlHttpRequest.responseText;
            alert(b);
            if (b=="true"){
                    alert(" 用户名已经存在 ");
            }else{
                            alert(" 用户名可以使用 ");
            }
          }
        }
    }
            function pass(){
            var pass = false;
            if( document.forms[0].elements["item.uname"].value =="" ){
                    alert(" 用户名不能为空 ");
                    pass= false;
            }else if(document.forms[0].elements["item.upass"].value == ""){
                    alert(" 密码不能为空 ");
                    pass = false;
            } else if(document.forms[0].upass.value != document.forms[0].elements["item.upass"].value){
                    alert(" 两次密码不一样 ");
                    pass = false;
            } else {
                    pass = true;
            }
            return pass;
        }
</script>
</HEAD>
```

```html
<BODY BGCOLOR=#FFFFFF LEFTMARGIN=0 TOPMARGIN=0 MARGINWIDTH=0 MARGINHEIGHT=0>

  <table width="780" border="0" align="center" cellpadding="0" cellspacing="0">
   <tr>
     <td colspan="5"><img src="images/top.jpg" width="780" height="213"></td>
   </tr>
   <tr>
     <td colspan="5"><img src="images/middle1.jpg" width="780" height="47"></td>
   </tr>
   <tr>
     <td width="38" background="images/middle2.jpg"> </td>
     <td width="172">
       <table align="center">
            <tr>
              <td><a href="#"> 返回首页 </a></td>
            </tr>
            <tr>
              <td><a href="#"> 用户登陆 </a></td>
            </tr>
       </table>
     </td>
     <td width="35" background="images/layout_24.gif"> </td>
     <td width="495">
       <html:form action="user.do?operate=doReg" method="post" onsubmit="return pass();">
              <table align="center">
                  <tr>
                       <td> 用户注册: </td>
                       <td> </td>
                  </tr>
                  <tr>
                       <td colspan="2"><hr/></td>
                  </tr>
                  <tr>
                       <td> 用户名: </td>
```

```
                            <td><html:text   property="item.uname"   onblur="return
checkUserExists(this);" /></td>
                        <tr>
                            <td>密码：</td>
                            <td><html:password property="item.upass" /></td>
                        <tr>
                        <tr>
                            <td>重复密码：</td>
                            <td><input type="password" name="upass"></td>
                        <tr>
                            <td><input   type="submit"   value=" 注 册 "   class="bt-
n"> </td>
                            <td><input type="reset" value=" 重置 " class="btn"></
td>
                        </tr>
                </table>
            </html:form>
            </td>
            <td width="40" background="images/middle4.jpg"> </td>
        </tr>
        <tr>
            <td colspan="5"><img src="images/bottom.jpg" width="780" height="82"></td>
        </tr>
    </table>
    <P align="center">2007 Beijing Aptech Beida Jade Bird Information Technology
Co.,Ltd 版权所有 </P>
    <br/>
    </BODY>
    </HTML>
```

该页面中用户名检测的功能使用了我们前面学习的 Ajax 技术。

这个成熟的三层架构 +MVC 的集成框架在稳定性与生产效率上获得了极佳的平衡，是很多企业做应用开发的首选。

11.4 小结

✓ Spring 为 Hibernate 开发提供了良好的支持，可以大大简化 Hibernate 的编码。

第 11 章 Spring 与 Struts、Hibernate 的集成

✓ 在 Spring 配置文件定义 DataSource()、SessionFactory() 再注入 DAO 是开发中常用的做法。
✓ Spring 依赖注入在项目中用于管理程序间的依赖，使我们更容易做到面向接口编程。
✓ Spring 可以采用 Struts 插件的方式与 Struts 轻松集成。
✓ Spring 支持声明式事务，可以以不写一行程序代码的方式给系统增加事务支持。Spring 2.5 的方式更为简洁、易懂。

11.5 英语角

transaction	事务
proxy	代理
target	目标
advice	通知
pointcut	切点
expression	表达式

11.6 作业

整合 SSH 框架，完成房屋信息管理系统的房屋信息查询、添加、修改和删除。

11.7 思考题

三大框架整合项目如何搭建？

11.8 学员回顾内容

1. Spring 与 Hibernate 的集成。
2. Spring 与 Struts 的集成。
3. 声明式事务。

上机部分

第1章 Java EE 概述

本阶段目标

完成本章内容后,你将能够:
◇ 了解 Ajax 的初步。

本阶段给出的步骤全面详细,请学员按照给出的上机步骤独立完成上机练习,以达到要求的学习目标。请认真完成下列步骤。

1.1 指导(1小时10分钟)

本章我们将用一个实际的案例介绍 BEA WebLogic 服务器的安装和配置。作为 BEA WebLogic Enterprise Platform 产品中最重要的一部分,WebLogic 服务器为开发和部署 Java EE 的应用程序提供了强有力的支持。

BEA WebLogic 服务器提供所有核心应用服务器应该提供的功能和服务,例如:
- 负载平衡(Load balancing)
- 容错(Fault tolerance)
- Web 服务(Web Services)
- 与大型机结合(Legacy integration)
- 事务管理(Transaction management)
- 安全(Security)
- 多线程(Multi-threading)
- 持久性(Persistence)
- 数据库连接(Database connectivity)
- 资源池化(Resource pooling)

1.2 练习

下面我们开始介绍 BEA WebLogic 服务器的安装和部署。

● 安装前的准备（下载软件）

我们可以从 BEA 网站上（http://commerce.bea.com）下载 WebLogic Server 的评估版本,支持的操作系统有 Microsoft Windows NT/7/8、Linux、Unix 和 Solaris 等,也可以在此网址下载 WebLogic 服务器的旧版本。下载的时候需要免费注册成为有效的用户才可以下载,有两种下载安装方法:

Package installer——下载一个 EXE 文件,这个 EXE 是一个打包的文件,包括了所有的 WebLogic 服务器软件的各个部分。例如:WebLogic 服务器和相关的例子,WebLogic Workshop 和相关的例子。

Net installer——先下载一个 EXE 文件,然后运行该程序。该程序可以让你有选择的安装相关的部件,然后在线下载你选择安装的部件。这个 EXE 文件大概有 22MB。推荐大家使用 Package installer 下载方式。

● 系统需求

操作系统:可以在 Windows NT/7/8、Unix、Linux 等操作系统安装。

硬盘空间:大概需要 650 MB。

内存:最少 256 MB,推荐使用 512 MB 或更多的内存。

管理员权限:如果你想在 Windows 操作系统下,把 WebLogic 服务器以服务形式运行,那么你需要有管理员的权限进行这项设置。

许可协议（License）:WebLogic Server 的运行不能没有相应的许可协议。当你安装 WebLogic Server 后,安装程序自动的创建了一个可以评估使用的许可协议,该许可协议可以最大有 5 个并发的客户连接、15 个数据库连接,一年的使用有效期。

● 安装 WebLogic 服务器

点击下载的 EXE 文件,出现 BEA Systems Installer 窗口,指示安装程序正在解压相关的安装文件。显示 Welcome 窗口,指示在安装的过程中任何时间都可以点 Exit 按钮退出安装。点 Next 按钮继续下一步。如图 1-1 所示。

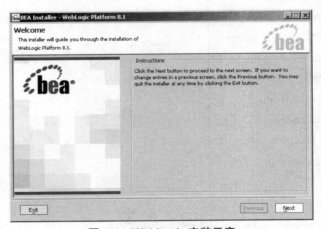

图 1-1　WebLogic 安装示意

如图 1-2 所示,显示 BEA Software License Agreement 窗口,选中 Yes,点 Next 按钮继续。

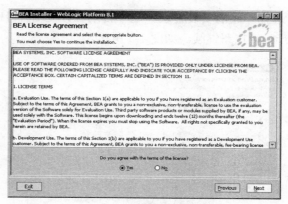

图 1-2　接受协议

选择 BEA Home 目录,如果你已经安装过 BEA 的产品,选择"Use an existing BEA Home",否则安装程序会默认选择"Create a new BEA Home"选项。在 Windows NT/7/8 操作系统下,默认的 BEA Home 目录是 C:\bea。如图 1-3 所示。

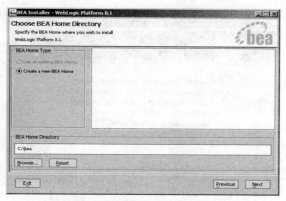

图 1-3　选择 BEA 主目录

然后出现选择安装类型的窗口,选择"Typical Installation",点 Next 按钮继续。如图 1-4 所示。

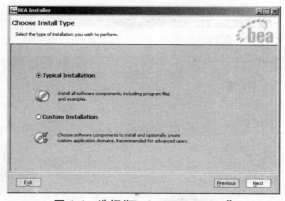

图 1-4　选择"Typical Installation"

下一个窗口让用户选择产品的安装目录。这个目录应该在 BEA Home 目录下，接受默认的目录（weblogic81），点 Next 按钮继续。如图 1-5 所示。

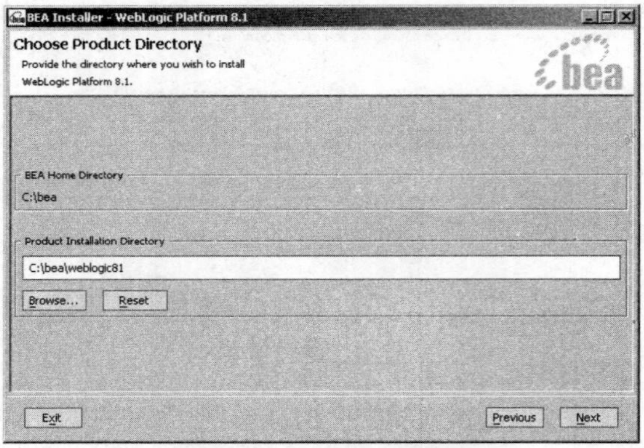

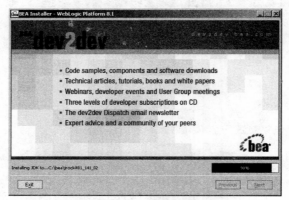

图 1-5　选择安装路径

安装程序开始往计算机中安装相应的文件。如图 1-6 所示。

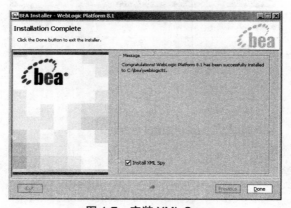

图 1-6　安装程序

当文件安装结束后，显示如下的窗口。指示可以继续选择安装额外的 XMLSPY 软件。根据自己的需求安装该软件。如图 1-7 所示。

图 1-7　安装 XML Spy

XMLSpy 软件是一个 XML 的工具软件,可以设计、编辑使用 XML、XML Schema、XSL/XSLT、SOAP、WSDL 和使用 Web Services 技术的应用程序,在 WebLogic 服务器中安装的这个 XML Spy 是一个特别版,有一些特殊的功能。通常,当安装完 WebLogic 服务器后,可以选择安装 XML Spy 。如果你当时你没有选择安装 XMLSpy,也可以运行 C:\bea\weblogic81\common\eval\xmlspy 目录下的 XMLSPYsetup.msi 文件进行安装。

● 配置 WebLogic 服务器

从开始菜单中运行 Configuration Wizard,配置一个 WebLogic 服务器的域(domain)。如图 1-8 所示。

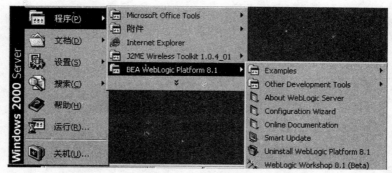

图 1-8 开始运行

一个域在 WebLogic 当中是一个基本的管理单元,它可以包含一个或多个运行着的 WebLogic 服务器,而这些 WebLogic 服务器和它们相关的资源都由一个单一的管理服务器(Administration Server)进行管理。一个最小的域可以只包含一个 WebLogic 服务器,这个 WebLogic 服务器既是管理服务器又是被管理的服务器(Managed Server),这种域通常在开发阶段被采用,但是在实际的产品运行阶段一个域通常会有单一的管理服务器和若干个被管理的服务器。

在弹出的窗口中,选择创建一个新的设置。如图 1-9 所示。

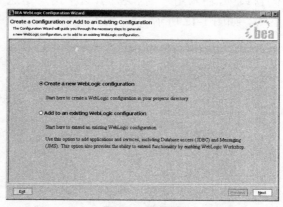

图 1-9 设置域

选择一个 domain 的设置模板,一个 domain 的设置模板提供了一个基本的 domain 的结构,你可以对这个设置模板进行相应修改,以满足你的需要。选择 Basic WebLogic Server Domain，点 Next 按钮继续 。如图 1-10 所示。

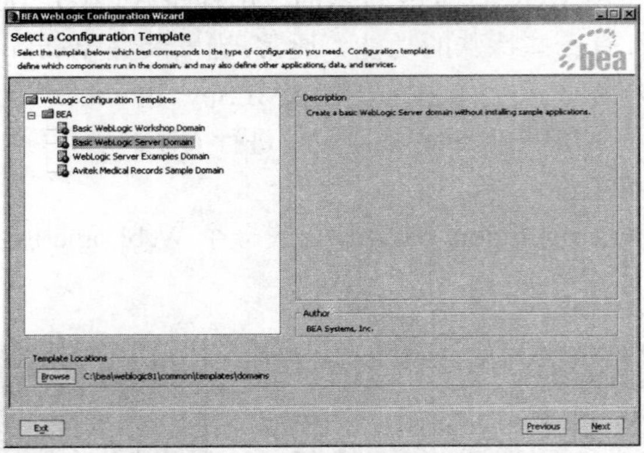

图 1-10　选择模板

在接下来的窗口中，选择 Express，点 Next 继续。如图 1-11 所示。

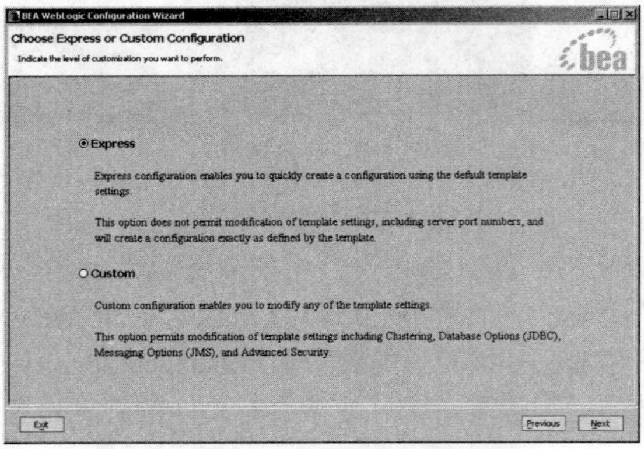

图 1-11　选择安装方式

在设置用户名和密码窗口中，输入用户名和密码，这个用户是这个 domain 的管理员，点 Next 按钮和确定密码后继续。如图 1-12 所示。

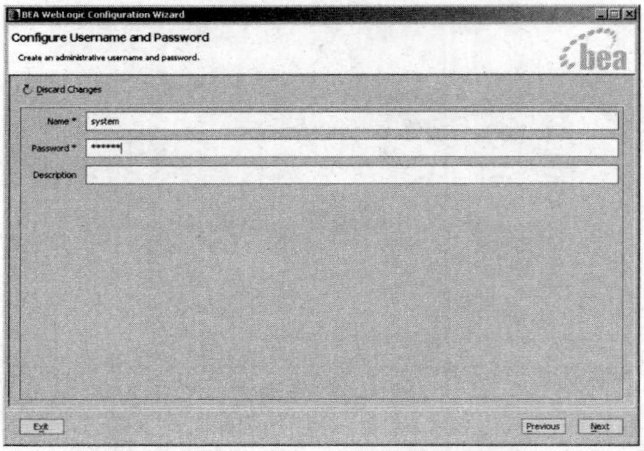

图 1-12　输入用户名和密码

在创建 WebLogic 设置窗口,点 Create 按钮。再点 Done 按钮完成 domain 设置。

domain 的名字默认 mydomain,这个 domain 中包含一个名字为 myserver 的单一服务器,这个服务器的根目录是 C:\bea\user_projects\mydomain。如图 1-13 所示。

图 1-13　创建设置

配置完成以后,我们就可以从开始菜单中启动这个服务器。如图 1-14 所示。

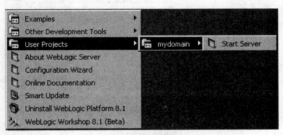

图 1-14　启动服务器

图 1-15 显示的是这个 WebLogic 服务器运行时启动的 DOS 窗口。这个 DOS 窗口会显示一些 WebLogic 服务器运行过程的输出信息,对我们在 WebLogic 服务器中调试自己的程序非常有帮助。

图 1-15　服务器运行中

为了验证这个 domain 是否设置正确,我们可以通过访问 WebLogic 服务器的管理界面进行验证。输入 http://localhost:7001/console/ 就可以访问 WebLogic 服务器的管理界面。如图 1-16 所示。输入在配置 domain 时候所设置的管理员的用户名和密码,输入正确后,就会显示如图 1-17 所示的页面。我们对 WebLogic 服务器的设置都是通过这个管理界面完成的。

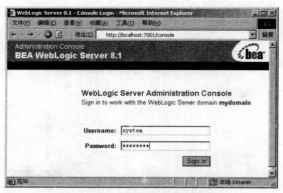

图 1-16 输入用户名和密码

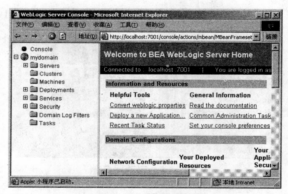

图 1-17 管理界面

- 修改环境变量

右击桌面上"我的电脑"的图标,选中属性,在弹出的窗口中,选中"高级"选项卡,在"高级"选项卡中点"环境变量",先把用户变量中 TEMP 和 TMP 的参数修改成和系统变量 TEMP 和 TMP 一样的数值,例如:C:\WINNT\TEMP。如图 1-18 所示。

图 1-18　修改 TEMP 和 TMP 变量

在系统变量中，点新建按钮，增加一个 WL_HOME 的参数，数值为你安装的 BEA WebLogic 服务器的目录，例如：c:\bea\weblogic81\server。如图 1-19 所示。

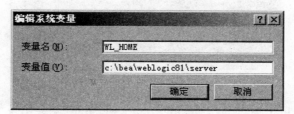

图 1-19　增加 WL_HOME 参数

如图 1-20 所示，修改系统变量中的 CLASSPATH 参数，加上 webloigic.jar 文件的路径。如果计算机中的系统变量中没有 CLASSPATH 参数，则添加一个 CLASSPATH 的参数，数值为 dt.jar、tools.jar 和 webloigic.jar 文件的路径。例如：c:\bea\jdk1.6\lib\dt.jar;c:\jdk141_02\lib\tools.jar;%WL_HOME%\lib\weblogic.jar。

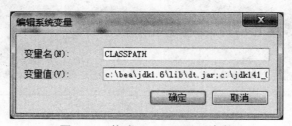

图 1-20　修改 CLASSPATH 参数

如图 1-21 所示修改系统变量中的 path 参数，加上 jdk 中 bin 目录的路径。例如：c:\bea\jdk1.6\bin。

图 1-21 修改 path 参数

- 验证环境变量的修改

选中"开始"菜单中的运行,在弹出的"运行"菜单中输入 cmd,在弹出的 DOS 窗口中运行 javac,就会出现如图 1-22 所示信息;如果没有出现如图 1-22 所示的信息,则系统变量中的 path 参数没有设置正确。

图 1-22 验证环境变量

在弹出的 DOS 窗口中运行 java weblogic.ejbc –version,就会出现如图 1-23 所示信息;如果没有出现图 1-23 的信息,可能是系统变量中的 CLASSPATH 参数没有设置正确。

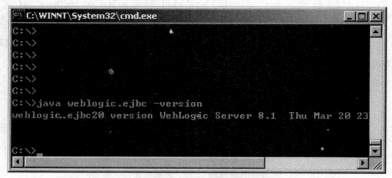

图 1-23 验证环境变量

1.3 作业

安装和配置 BEA WebLogic 服务器，并测试。

第 2 章 使用 Ajax 改进用户体验

本阶段目标

完成本章内容后,你将能够:
✧ 了解 Ajax 的初步。
✧ 掌握 Ajax 的实现步骤。

本阶段给出的步骤全面详细,请学员按照给出的上机步骤独立完成上机练习,以达到要求的学习目标。请认真完成下列步骤。

2.1 指导 (1 小时 10 分钟)

本次上机的任务是完成一个用户登录的示例,与以往不同,由于采用了异步交互方式,所以在显示验证结果的提示信息时没有页面刷新。

2.2 Ajax 的实现方式完成用户登录

现在,我们使用 Ajax 的方式实现异步交互的效果,图 2-1 显示了登录成功后显示欢迎信息的界面。可以看到提示信息显示的窗口后面的页面并不是空白的,在页面状态栏中也看不到页面刷新显示的进度条。

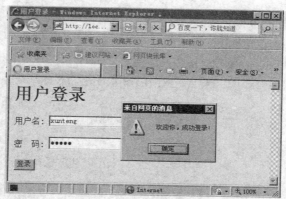

图 2-1 登陆成功

图 2-2 所示页面为登录失败后的提示信息窗口的效果。在该页面效果中同样也看不到页面在进行刷新时的效果。

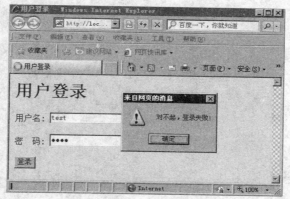

图 2-2 登录失败

下面,我们来看 Ajax 的异步交互方式是如何实现的。

第一步,在 MyEclipse 中新建一个 Web 项目,将该项目命名为 AjaxLogin。该项目的目录结构上包含一个登录页面 Login.jsp,以及进行登录验证处理的 Servlet,类名为 Login。登录页面 Login.jsp 中加入了与 Ajax 技术相关的程序代码,而 Servlet 处理完登录验证后返回给客户端的是 XML 格式的数据。Login.jsp 的源代码如示例代码 2-1 所示。

```
示例代码 2-1：Login.jsp
<%@ page language="java" pageEncoding="gb2312"%>
<html>
    <head>
            <title> 用户登录 </title>
    </head>
    <script type="text/javascript">
            var xmlHttp;
```

```
// 创建 XMLHttpRequest()
function createXMLHttpRequset(){
        if(window.XMLHttpRequest){
                xmlHttp = new XMLHttpRequest();
        }
        else if(window.ActiveXObject){
                try{
                        xmlHttp = new ActiveXObject("Msxml2.XMLHTTP");
                }catch(e){
                        try{
                                xmlHttp = new ActiveXObject("Microsoft.XMLHTTP");
                        }catch(e){}
                }
        }
}
function createQueryString(){
        var userName=document.myForm.userName.value;
                var password=document.myForm.password.value;
        var queryStr = "userName="+userName+"&password="+password;
        return queryStr;
}
function doRequest(){
        createXMLHttpRequset();
        var url="login?"+createQueryString();

        xmlHttp.open("GET",url,true);
        xmlHttp.onreadystatechange=handleResponse;
        xmlHttp.setRequestHeader("Content-Type"
          ,"application/x-www-form-urlencoded");
        xmlHttp.send(null);
}
 function handleResponse(){

  if(xmlHttp.readyState==4){
    if(xmlHttp.status==200){
      var xml=xmlHttp.responseXML;
      var result=xml.getElementsByTagName("result")[0]
                  .firstChild.nodeValue;
```

```html
            alert(result);
          }else{
            alert(" 您请求的页面有异常 ");
          }
        }
      }
    </script>
    <body bgcolor="C8F8F8">
          <h1>
                用户登录
          </h1>
          <form action="" name="myForm">
              用户名：
              <input type="text" name="userName">
              <br>
              <br>
              密    码：
              <input type="password" name="password">
              <br>
              <br>
              <input type="submit" value=" 登录 " onclick="doRequest()">
          </form>
    </body>
</html>
```

从上面的代码，我们可以看到，当用户输入了用户名和密码，并点击"登录"按钮时将会调用 JavaScript 函数 doRequest()。在该函数中将使用 Ajax 技术来提交请求，并接收服务器端请求处理后的结果数据，然后进行显示。

第二步，创建服务器端 Servlet 处理程序。注意：该 Servlet 完成验证后返回的是 XML 格式的数据。具体的代码如示例代码 2-2 所示。

示例代码 2-2：LoginServlet.java

```java
package xtgj.servlets;


import java.io.IOException;
import java.io.PrintWriter;
```

```java
import javax.servlet.ServletException;
import javax.servlet.http.HttpServlet;
import javax.servlet.http.HttpServletRequest;
import javax.servlet.http.HttpServletResponse;

public class LoginServlet extends HttpServlet {
    public void handleRequest(HttpServletRequest request,
                HttpServletResponse response, String method)
                    throws ServletException, IOException {
        String userName = request.getParameter("userName");
        String password = request.getParameter("password");
        response.setContentType("text/xml;charset=utf-8");
        response.setHeader("Cache-Control", "no-cache");
        PrintWriter out = response.getWriter();
        String xml = "<response>";
        if (userName.equals("xunteng") && password.equals("guoji")) {
                xml += "<result>欢迎你,成功登录!</result>";
        } else {
                xml += "<result>对不起,登录失败!</result>";
        }
        xml += "</response>";
        System.out.println(xml);
        out.println(xml);
        out.close();
    }

    public void doGet(HttpServletRequest request, HttpServletResponse response)
                throws ServletException, IOException {
        handleRequest(request, response, "GET");
    }

    public void doPost(HttpServletRequest request, HttpServletResponse response)
                throws ServletException, IOException {
        handleRequest(request, response, "POST");
    }
}
```

从本案例中的代码可以看到,无论视图层的页面还是控制层的 Servlet 程序都发生了改变,这主要是因为采用了 Ajax 技术提供的 XMLHttpRequest 对象,由它来负责向服务器提交请求以及接受从服务器发回的响应数据。

第三步，配置 Servlet，web.xml 源代码如示例代码 2-3 所示。

示例代码 2-3：web.xml
```xml
<?xml version="1.0" encoding="UTF-8"?>
<web-app version="3.0"
    xmlns="http://java.sun.com/xml/ns/javaee"
    xmlns:xsi="http://www.w3.org/2001/XMLSchema-instance"
    xsi:schemaLocation="http://java.sun.com/xml/ns/javaee
    http://java.sun.com/xml/ns/javaee/web-app_3_0.xsd">

    <servlet>
      <servlet-name>Login</servlet-name>
      <servlet-class>xtgj.servlets.LoginServlet</servlet-class>
    </servlet>

    <servlet-mapping>
      <servlet-name>Login</servlet-name>
      <url-pattern>/Login</url-pattern>
    </servlet-mapping>
    <welcome-file-list>
      <welcome-file>index.jsp</welcome-file>
    </welcome-file-list>
</web-app>
```

在创建完项目文件之后，将该项目部署到 Tomcat 服务器，并启动 Tomcat，然后在浏览器地址栏中输入如下的地址查看运行效果：http://localhost:8080/AjaxLogin/login.jsp。

2.2.1 Ajax 的实现方式完成词典查询

本案例将使用 Ajax 的实现方式完成词典查询，当用户在文本框中输入单词，并点击"词典查询"按钮，将使用 XMLHttpRequest 向服务器发送一个异步请求，服务器端的一个 Servlet 处理该请求，并使用简单文本方式返还响应。该响应数据包含了 HTML 内容。客户端则使用 responseText 属性获取响应数据，并利用元素的 innerHTML 属性将其显示。具体步骤如下：

第一步，创建 Web 项目，命名为 AjaxDict，在该项目中创建名为"dict.html"的文件，代码如示例代码 2-4 所示。

示例代码 2-4：dict.html

```html
<html>
    <head>
            <title>dict.html</title>
            <script type="text/javascript">
    var xmlHttp;
    // 创建 XMLHttpRequest 对象函数
    var xmlHttp;
    function createHttpRequest() {
            if (window.XMLHttpRequest) {
                    xmlHttp = new XMLHttpRequest();
            } else if (window.ActiveXObject) {
                    try {
                            xmlHttp = new ActiveXObject("Msxml2.XMLHttp");
                    } catch (e) {
                            try {
                                    xmlHttp = new ActiveXObject("Microsoft.XMLHTTP");
                            } catch (e) {
                            }
                    }
            }
    }
    // 发送请求函数
    function doRequest() {
            createHttpRequest();
            // 获取用户输入的单词
            var word = document.getElementById("word").value;
            if (word == "")
                    return;
            // 处理请求的 Servlet 的 url
            var url = "Search";
            // 请求体
            var queryContent = "key=" + word;
            xmlHttp.onreadystatechange = handleResponse;
            // 以 post 方式发送请求
            xmlHttp.open("POST", url, true);
            //POST 方式需要设置 XMLHttpRequest 对象的 Content-Type 首部
```

```
                    xmlHttp.setRequestHeader("Content-Type",
                            "application/x-www-form-urlencoded");
                    // 发送请求，参数为发送的请求体内容
                    xmlHttp.send(queryContent);
            }
            // 处理响应的函数
            function handleResponse() {
                    if (xmlHttp.readyState == 4) {
                            if (xmlHttp.status == 200) {
                                    // 获取 div 结点
                                    var resultDiv = document.getElementById("result");
                                    // 用 div 结点的 innerHTML 属性来显示包含 HTML 内容的响应数据
                                    resultDiv.innerHTML = xmlHttp.responseText;
                            } else {
                                    alert(" 您请求的页面有异常 ");
                            }
                    }
            }
        </script>
    </head>

    <body>
            <input id="word" type="text" size="30">
              
            <input type="button" value=" 词典查询 " onclick="doRequest()">
            <br>
            <div id="result"></div>
    </body>
</html>
```

第二步，创建服务器端处理程序，相应的 Servlet 源代码如示例代码 2-5 所示。

示例代码 2-5：Search.java

```java
package servlets;

import java.io.IOException;
import java.io.PrintWriter;
```

```java
import javax.servlet.ServletException;
import javax.servlet.http.HttpServlet;
import javax.servlet.http.HttpServletRequest;
import javax.servlet.http.HttpServletResponse;

public class Search extends HttpServlet{

    public void doPost(HttpServletRequest request,HttpServletResponse response)
                throws ServletException,IOException{
        String word = request.getParameter("key");
        // 拼接一个 html 表格作为相应内容，将其表示为一个字符串
        System.out.println(word);
        StringBuffer answer = new StringBuffer("<table border='1' cellpadding='5'>");
        if(word.equalsIgnoreCase("java")){
            answer.append("<tr><td>")
                .append("n. 爪哇岛（印度尼西亚）")
                .append("</td></tr>")
                .append("<tr><td>")
                .append("n. 一种咖啡（肯尼亚）")
                .append("</td></tr>")
                .append("<tr><td>")
                .append("n. 一种优秀的,面向对象得编程语言（学习ing....)")
                .append("</td></tr>");
        }
        else{
            answer.append("<tr><td>")
                .append(" 你输入的单词没找到 ")
                .append("</td></tr>");
        }
        answer.append("</table>");

        // 设置相应内容类型为简单文本 text/palin
        response.setContentType("text/plain;charset=UTF-8");
        response.setHeader("Cache-Control","no-cache");
        PrintWriter out = response.getWriter();
```

```
        // 发送响应字符串
        out.println(answer.toString());
        out.close();
    }
}
```

第三步,配置 Servlet,web.xml 源代码如示例代码 2-6 所示。

示例代码 2-6：web.xml
```
<?xml version="1.0" encoding="UTF-8"?>
<web-app version="3.0"
    xmlns="http://java.sun.com/xml/ns/javaee"
    xmlns:xsi="http://www.w3.org/2001/XMLSchema-instance"
    xsi:schemaLocation="http://java.sun.com/xml/ns/javaee
    http://java.sun.com/xml/ns/javaee/web-app_3_0.xsd">
<servlet>
  <servlet-name>Search</servlet-name>
  <servlet-class>servlets.Search</servlet-class>
 </servlet>
 <servlet-mapping>
  <servlet-name>Search</servlet-name>
  <url-pattern>/Search</url-pattern>
 </servlet-mapping>
 <welcome-file-list>
  <welcome-file>index.jsp</welcome-file>
 </welcome-file-list>
</web-app>
```

在创建完项目文件之后,将该项目部署到 Tomcat 服务器,并启动 Tomcat,然后在浏览器地址栏中输入如下的地址查看运行效果：http://localhost:8080/AjaxDict/dict.html。如图 2-3 所示。

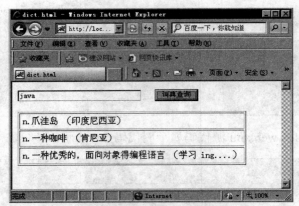

图 2-3 Ajax 词典查询

2.3 作业

1. 创建一个智力测试程序。要求创建一个带有表单的 html 页面，表单中显示 3 个选择题，每题有 4 个选项，使用单选按钮供用户选择。当用户选择完毕按"提交"按钮后，由 JavaScript 计算用户的答题成绩，并使用消息框显示。

2. 练习使用 CSS 样式表控制页面元素的显示方式。

第 3 章　Ajax 实战技巧

本阶段目标

完成本章内容后，你将能够：
- 了解 Ajax 的使用场合。
- 掌握 Ajax 的实现技巧。

本阶段给出的步骤全面详细，请学员按照给出的上机步骤独立完成上机练习，以达到要求的学习目标。请认真完成下列步骤。

3.1　指导(1 小时 10 分钟)

3.1.1　用户注册

本指导练习中，我们来实现一个用户注册的实例。该实例主要用于实现用户注册时检测用户名是否已经存在。其实现效果如图 3-1 至图 3-4 所示。

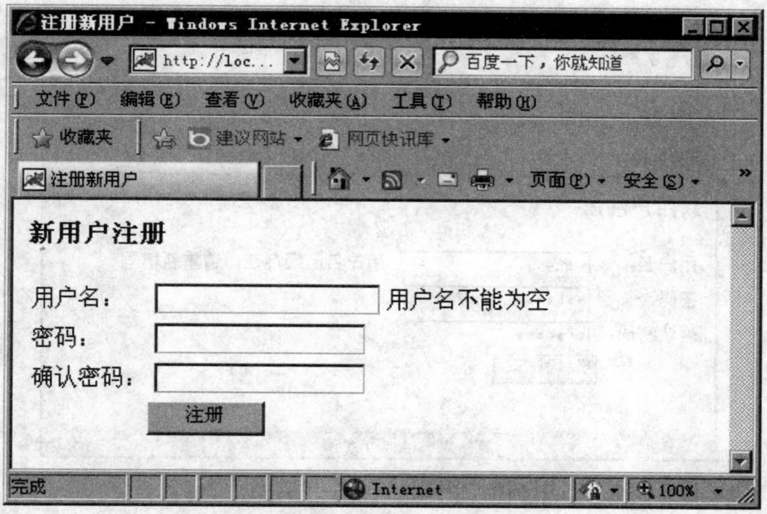

图 3-1　检测用户名是否为空

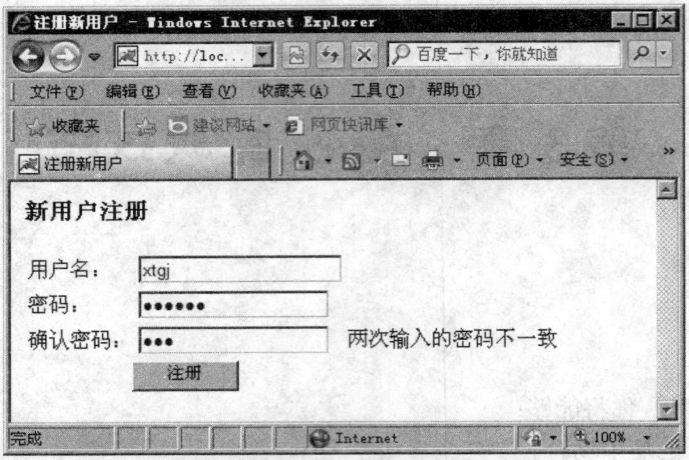

图 3-2 检测两次输入的密码是否一致

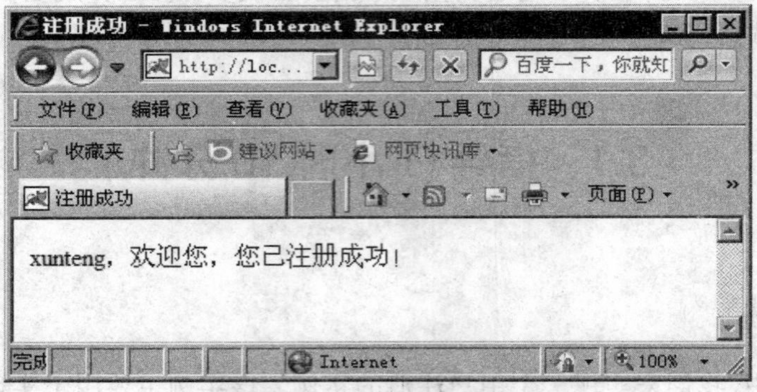

图 3-3 注册成功的页面

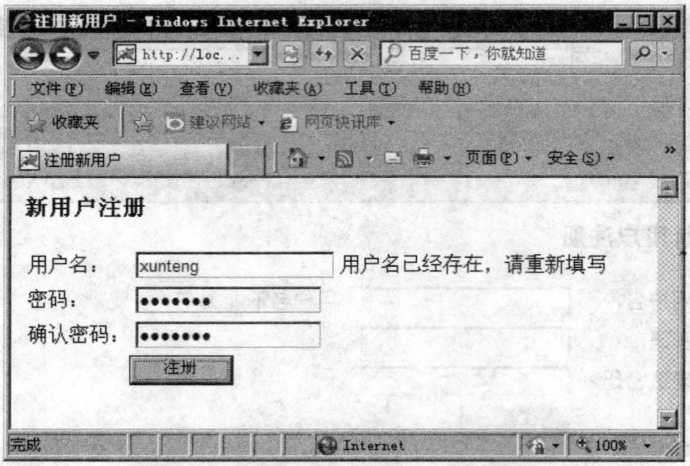

图 3-4 检测用户名是否已经存在

当用户单击"注册"按钮后,将对用户填写的信息进行验证,如果用户没有填写用户名、密码,或者两次输入的密码不一致,都会在输入框的右边显示提示信息。此外,当用户所填写的用户名已经被占用时，Ajax 所调用的服务器处理程序同样会返回提示信息。如果注册成功，则跳转到注册成功页面显示欢迎信息。

本案例需要一个用户表 user，我们在 SqlServer 的 test 数据库执行如示例代码 3-1 所示的脚本。

示例代码 3-1：test.sql

```sql
create table user_table(
id int identity primary key,
username varchar(20) not null,
password varchar(20) not null
)
```

第一步，本案例所使用的项目环境仍然为理论部分的"Chapter03"。

第二步，创建该项目需要使用的系统状态 Bean，User.java 用于封装用户的信息。代码如示例代码 3-2 所示。

示例代码 3-2：User.java

```java
package com.xtgj.j2ee.chapter03.user.beans;

public class User {
    private int id;
    private String userName;
    private String password;
    public int getId() {
            return id;
    }
    public void setId(int id) {
            this.id = id;
    }
    public String getUserName() {
            return userName;
    }
    public void setUserName(String userName) {
            this.userName = userName;
    }
    public String getPassword() {
            return password;
    }
    public void setPassword(String password) {
```

```
            this.password = password;
        }
}
```

第三步,创建业务逻辑 Bean,DBOperator.java 实现数据库的操作,包括检查用户名是否被占用,以及向 user 表插入新用户的注册信息。代码如示例代码 3-3 所示。

示例代码 3-3:DBOperator.java

```java
package com.xtgj.j2ee.chapter03.user.beans;

import java.sql.Connection;
import java.sql.PreparedStatement;
import java.sql.ResultSet;
import java.sql.Statement;
import com.xtgj.j2ee.chapter03.db.DBUtil;

public class DBOperator {
    // 保存用户注册信息
    public void save(User user) throws Exception {
        Connection cn = null;
        PreparedStatement pst = null;
        ResultSet rs = null;
        try {
            cn = getConnection();
            cn.setAutoCommit(false);
            String sql = "insert into user_table(userName,password) values(?,?)";
            pst = cn.prepareStatement(sql, Statement.RETURN_GENERATED_KEYS);

            pst.setString(1, user.getUserName());
            pst.setString(2, user.getPassword());

            pst.executeUpdate();
            // 获取新插入记录的自增 id
            rs = pst.getGeneratedKeys();
            rs.next();
            int id = rs.getInt(1);
            user.setId(id);
            cn.commit();
```

```java
        } catch (Exception e) {
            e.printStackTrace();
            cn.rollback();
            throw e;
        } finally {
            try {
                rs.close();
            } catch (Exception e) {
            }
            try {
                pst.close();
            } catch (Exception e) {
            }
            try {
                cn.close();
            } catch (Exception e) {
            }
        }
    }
    // 检查用户是否被占用
    public boolean existUserName(String userName) throws Exception {
        boolean exist = false;
        Connection cn = getConnection();
        String sql = "select count(*) from user_table where userName=?";
        PreparedStatement pst = cn.prepareStatement(sql);
        pst.setString(1, userName);
        ResultSet rs = pst.executeQuery();
        rs.next();
        if (rs.getInt(1) == 1) {
            exist = true;
        }
        rs.close();
        pst.close();
        cn.close();
        return exist;
    }
    // 获取数据库连接对象
    private Connection getConnection() throws Exception {
```

```
                return DBUtil.getConn();
    }
}
```

第四步，创建用户注册界面，命名为 register.jsp。代码如示例代码 3-4 所示。

示例代码 3-4：register.jsp

```jsp
<%@ page language="java" import="java.util.*" pageEncoding="UTF-8"%>
<html>
 <head>
  <title>注册新用户</title>
  <script type="text/javascript">
    var xmlHttp;
    // 创建 XMLHttpRequest 对象
    function createXMLHttpRequest(){
            if(window.XMLHttpRequest){
                    xmlHttp = new XMLHttpRequest();
            }else if(window.ActiveXObject){
                    try{
                            xmlHttp = new ActiveXObject("Msxml2.XMLHTTP");
                    }catch(e){
                            try{
                            xmlHttp = new ActiveXObject("Microsoft.XMLHTTP");
                            }catch(e){}
                    }
            }
    }
    // 发送请求函数
    function sendRequest(queryStr){
            createXMLHttpRequest();
            var url="Register";
            xmlHttp.open("POST",url,true);
            xmlHttp.onreadystatechange = handleResponse;
            xmlHttp.setRequestHeader("Content-Type"
               ,"application/x-www-form-urlencoded");
            xmlHttp.send(queryStr);
    }
    // 处理响应函数
    function handleResponse(){
```

```js
            if(xmlHttp.readyState==4){
                if(xmlHttp.status==200){
                    var xml = xmlHttp.responseXML;
                    var res = xml.getElementsByTagName("result")[0]
                        .firstChild.nodeValue;
                    if(res =="true"){
                        window.location="regOK.jsp";
                    }else{
                        var msg = xml.getElementsByTagName("message")[0]
                        .firstChild.nodeValue;
                        var tdMsg = document.getElementById("msg1");
                        tdMsg.innerHTML = msg;
                    }
                }else{
                    alert(" 您请求的页面有异常！ ");
                }
            }
        }

// 注册
function register(){
    var userName = document.regForm.userName.value;
    var password = document.regForm.password.value;
    var repeatPwd = document.regForm.repeatPwd.value;
    document.getElementById("msg1").innerHTML="";
    document.getElementById("msg2").innerHTML="";
    document.getElementById("msg3").innerHTML="";

    if(userName==""){
        document.getElementById("msg1")
            .innerHTML=" 用户名不能为空 ";
        return;
    }
    if(password==""){
        document.getElementById("msg2")
            .innerHTML=" 密码不能为空 ";
        return;
```

```
                    }
                    if(repeatPwd!=password){
                            document.getElementById("msg3")
                                .innerHTML=" 两次输入的密码不一致 ";
                            return;
                    }
                    var queryStr="userName="+userName+"&password="+password;
                    sendRequest(queryStr);
            }
        </script>
    </head>

    <body>
        <h3> 新用户注册 </h3>
        <form action="" name="regForm">
            <table>
                <tr>
                    <td> 用户名：</td>
                    <td><input type="text" name="userName"></td>
                    <td id="msg1"></td>
                </tr>
                <tr>
                    <td> 密码：</td>
                    <td><input type="password" name="password"></td>
                    <td id="msg2"></td>
                </tr>
                <tr>
                    <td> 确认密码：</td>
                    <td><input type="password" name="repeatPwd"></td>
                    <td id="msg3"></td>
                </tr>
                <tr>
                    <td colspan="2" align="center">
                        <input type="button" value=" 注册 "
                            onclick="register();">
                    </td>
                </tr>
            </table>
```

```
    </form>
  </body>
</html>
```

当用户单击"注册"按钮后,将调用 register() 函数。在 register() 函数中首先检测用户填写数据是否非空,以及两次输入密码的一致性。如果通过验证,则调用 sendRequest() 函数将用户注册信息发送到服务器进行处理。服务器端程序将会检测用户名是否被占用,如果未占用则将注册信息保存至数据库,同时将相应的响应信息发回至客户端。

handleResponse() 函数处理服务器端的响应信息,如果用户名已被占用则显示相应的信息,提示用户重新填写。如果注册成功则跳转到注册成功页面 regOK.jsp。

第五步,创建服务器端 Servlet 处理程序,类名为 Register.java。其代码如示例代码 3-5 所示。

示例代码 3-5:Register.java

```java
package com.xtgj.j2ee.chapter03.servlets;

import java.io.IOException;
import java.io.PrintWriter;

import javax.servlet.ServletException;
import javax.servlet.http.HttpServlet;
import javax.servlet.http.HttpServletRequest;
import javax.servlet.http.HttpServletResponse;

import com.xtgj.j2ee.chapter03.user.beans.DBOperator;
import com.xtgj.j2ee.chapter03.user.beans.User;

public class Register extends HttpServlet {

    public void doPost(HttpServletRequest request, HttpServletResponse response)
            throws ServletException, IOException {

        request.setCharacterEncoding("UTF-8");
        // 获取用户填写的注册信息
        String userName = request.getParameter("userName");
        String password = request.getParameter("password");
        // 检测用户名是否被占用
        DBOperator db = new DBOperator();
        boolean exist = false;
```

```java
            try {
                    exist = db.existUserName(userName);
            }
            catch (Exception e) {
                    e.printStackTrace();
                    throw new ServletException(" 数据库错误 ");
            }
            // 设置相应格式和字符集
            response.setContentType("text/xml;charset=utf-8");
            response.setHeader("Cache-Control", "no-cache");
            PrintWriter out = response.getWriter();
            out.println("<response>");
            if(exist){
                    // 用户名已被占用的响应信息
                    out.println("<result>false</result>");
                    out.println("<message> 用户名已经存在,请重新填写 </message>");
            }else{
                    // 如果用户名未被占用,则保存用户注册信息
                    User user = new User();
                    user.setUserName(userName);
                    user.setPassword(password);
                    try {
                            db.save(user);
                    }
                    catch (Exception e) {
                            e.printStackTrace();
                            throw new ServletException(" 数据库错误 ");
                    }
                    // 将用户信息存放在 session 范围内
                    request.getSession().setAttribute("user", user);
                    // 输出响应信息表示用户注册成功
                    out.println("<result>true</result>");
            }
            out.println("</response>");
            out.close();
    }
}
```

第六步,创建注册成功页面 regOK.jsp,代码如示例代码 3-6 所示。

示例代码 3-6:regOK.jsp
```jsp
<%@ page language="java" import="java.util.*" pageEncoding="UTF-8"%>
<!DOCTYPE HTML PUBLIC "-//W3C//DTD HTML 4.01 Transitional//EN">
<html>
    <head>
            <title> 注册成功 </title>
    </head>
    <body>
            <%=((com.xtgj.j2ee.chapter03.user.beans.User) session.getAttribute("user")).getUserName()%>,欢迎您,您已注册成功!
            <br>
    </body>
</html>
```

从上面的程序中,我们可以看到,采用 Ajax 技术,页面不必完全刷新,这无疑会给注册用户带来更好的体验,同时也很好地解决了数据有效性问题。

3.1.2 创建自动刷新页面

在实际的 Web 应用中,像股票行情、热销商品排行、最新标题新闻等数据都是经常改变的,往往需要定期进行自动刷新,但不值得为这些数据修改完全刷新页面。如果只是为了改变一两个标题新闻就重画整个页面,这可能很让人扫兴,而且可能很难发现到底哪些是新内容。借助 Ajax 技术,可以实现对页面中局部区域的动态刷新,使得用户能够以更好的方式获得最新数据。

在这个练习中,我们来实现页面自动刷新的功能。这个例子的实现效果如图 3-5 所示。

在该页面中将根据数据库中存储的最新数据,定时更新热卖商品的排行信息,对这些数据的修改并不会导致整个页面的刷新。例如,联想笔记本的销量上涨,超过了华硕笔记本,那么页面局部刷新的结果如图 3-6 所示。

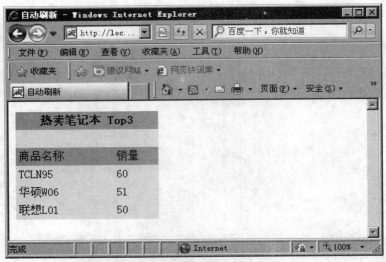

图 3-5 商品销售排行页面

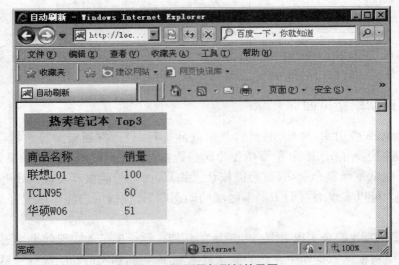

图 3-6 页面局部刷新效果图

本例需要一个商品信息表 product,我们在 SqlServer 的 test 数据库执行如示例代码 3-7 所示的脚本。

示例代码 3-7：test.sql

```
create table product(
id int identity primary key,
name varchar(20) not null,
```

```
price float,
saleCount int
);

insert into product(name,price,saleCount) values('联想 L01',6999,50);
insert into product(name,price,saleCount) values('戴尔 D05',56999,49);
insert into product(name,price,saleCount) values('华硕 W06',7599,51);
insert into product(name,price,saleCount) values('惠普 H28',7999,48);
insert into product(name,price,saleCount) values('TCLN95',5999,60);
```

第一步,使用上一个项目环境,并创建该项目需要使用的系统状态 Bean,Product.java 用于封装商品的信息。代码如示例代码 3-8 所示。

示例代码 3-8: Product.java

```java
package com.xtgj.j2ee.chapter03.pro.beans;

public class Product {
    private int id; // 商品编号
    private String name; // 商品名称
    private double price; // 价格
    private int saleCount; // 销售数量

    public int getId() {
        return id;
    }

    public void setId(int id) {
        this.id = id;
    }

    public String getName() {
        return name;
    }

    public void setName(String name) {
        this.name = name;
    }
```

```java
    public double getPrice() {
        return price;
    }

    public void setPrice(double price) {
        this.price = price;
    }

    public int getSaleCount() {
        return saleCount;
    }

    public void setSaleCount(int saleCount) {
        this.saleCount = saleCount;
    }

}
```

第二步,业务逻辑 Bean,DBOperator.java 用于实现数据库的操作,即查询销量前 n 名的商品。代码如示例代码 3-9 所示。

示例代码 3-9:DBOperator.java

```java
package com.xtgj.j2ee.chapter03.pro.beans;

import java.sql.Connection;
import java.sql.DriverManager;
import java.sql.PreparedStatement;
import java.sql.ResultSet;
import java.sql.Statement;
import java.util.ArrayList;
import java.util.List;

import com.xtgj.j2ee.chapter03.db.DBUtil;

public class DBOperator {
    // 获取数据库连接对象
    private Connection getConnection() throws Exception {
        return DBUtil.getConn();
```

```java
    }
    // 查询销售数量前 top 名的商品信息
    public List findHotProduct(int top) throws Exception {
        List list = new ArrayList();
        Connection cn = getConnection();
        Statement st = cn.createStatement();
        String sql = "select * from product order by saleCount desc";
        ResultSet rs = st.executeQuery(sql);
        int i = 0;
        while (rs.next()) {
            Product prod = new Product();
            prod.setId(rs.getInt(1));
            prod.setName(rs.getString(2));
            prod.setPrice(rs.getDouble(3));
            prod.setSaleCount(rs.getInt(4));
            list.add(prod);
            i++;
            if (i == top)
                break;
        }
        rs.close();
        st.close();
        cn.close();
        return list;
    }
}
```

第三步，创建热卖商品 JSP 页面，命名为 autoRefresh.jsp。代码如示例代码 3-10 所示。

示例代码 3-10：autoRefresh.jsp

```jsp
<%@ page language="java" import="java.util.*" pageEncoding="GBK"%>
<html>
 <head>
  <title> 自动刷新 </title>
 </head>
<script language="javascript">
    var xmlHttp;
```

```javascript
// 创建 XMLHttpRequest 对象
function createXMLHttpRequest(){
        if(window.XMLHttpRequest){
                xmlHttp = new XMLHttpRequest();
        }else if(window.ActiveXObject){
                try{
                        xmlHttp = new ActiveXObject("Msxml2.XMLHTTP");
                }catch(e){
                        try{
                                xmlHttp = new ActiveXObject("Microsoft.XMLHTTP");
                        }catch(e){}
                }
        }
}
// 发送请求
function sendRequest(){
        createXMLHttpRequest();
        var url = "Refresh";
        xmlHttp.open("GET",url,true);
        xmlHttp.onreadystatechange = handleResponse;// 指定响应函数
        xmlHttp.setRequestHeader("Content-Type"
            ,"application/x-www-form-urlencoded");
        xmlHttp.send(null);// 发送请求
}
// 处理返回响应信息函数
function handleResponse(){
        if(xmlHttp.readyState==4){// 判断对象状态
                if(xmlHttp.status==200){// 信息已经成功返回，开始处理信息
                        showHotList();
                        setTimeout("sendRequest()",5000);
                }else{
                        window.alert(" 您请求的页面有异常！");
                }
        }
}
// 显示热卖商品列表
function showHotList(){
        clearHotList();
```

```
            var hotList = document.getElementById("hotList");
            var xml = xmlHttp.responseXML;
            var ps = xml.getElementsByTagName("product");
            var row,cell,text;
                for(var i=0;i<ps.length;i++){

                    var name = ps[i].getElementsByTagName("name")[0]
                .firstChild.nodeValue;
                    var count = ps[i].getElementsByTagName("count")[0]
                .firstChild.nodeValue;
                    row = document.createElement("tr");
                    cell = document.createElement("td");
                    text = document.createTextNode(name);
                    cell.appendChild(text);
                    row.appendChild(cell);

                    cell = document.createElement("td");
                    text = document.createTextNode(count);
                    cell.appendChild(text);
                    row.appendChild(cell);
                    hotList.appendChild(row);
            }
            // 清除热卖商品列表
            function clearHotList(){
                    var hotList = document.getElementById("hotList");
                    while(hotList.childNodes.length>0){
                            hotList.removeChild(hotList.firstChild);
                    }
            }
        }
    </script>
    <body onload="sendRequest()">
<table cellspacing="0" cellpadding="4"
            width="200" bgcolor="#f5efe7">
    <tr bgcolor="#dbc2b0">
            <td align="center" colspan="2">
                <b>热卖笔记本 Top3</b>
        </td>
```

```html
            </tr>
            <tr>
                    <td> </td>
            </tr>
            <tr bgcolor="#dbc2b0">
                    <td width="140"> 商品名称 </td>
                    <td width="60"> 销量 </td>
            </tr>
            <tbody id="hotList"></tbody>
        </table>
    </body>
</html>
```

在该页面第一次加载时，将调用 sendRequest() 函数，该函数向服务器请求最热卖商品的信息。在处理响应信息的函数 handleResponse() 中，先调用 showHotList() 函数显示热卖商品信息，然后设置了定时器，即在 5000 毫秒后再次调用 sendRequest() 函数，重复执行上面的函数，从而实现页面的自动刷新。

第四步，创建服务器端 Servlet 处理程序，类名为 Refresh.java，代码如示例代码 3-11 所示。

示例代码 3-11：Refresh.java

```java
package com.xtgj.j2ee.chapter03.servlets;

import java.io.IOException;
import java.io.PrintWriter;
import java.util.List;
import javax.servlet.ServletException;
import javax.servlet.http.HttpServlet;
import javax.servlet.http.HttpServletRequest;
import javax.servlet.http.HttpServletResponse;

import com.xtgj.j2ee.chapter03.pro.beans.DBOperator;
import com.xtgj.j2ee.chapter03.pro.beans.Product;

public class Refresh extends HttpServlet {

    public void doGet(HttpServletRequest request, HttpServletResponse response)
            throws ServletException, IOException {
```

```java
// 查询销量排行前 3 名的商品信息
DBOperator db = new DBOperator();
List list = null;
try {
        list = db.findHotProduct(3);
} catch (Exception e) {
        throw new ServletException(" 数据库错误 ");
}
// 设置响应格式和字符集

// 以 XML 格式输出商品信息

response.setContentType("text/xml;charset=utf-8");
response.setHeader("Cache-Control", "no-cache");
PrintWriter out = response.getWriter();
out.println("<response>");
for (int i = 0; i < list.size(); i++) {
        Product p = (Product) list.get(i);
        out.println("<product>");
        out.println("<name>" + p.getName() + "</name>");
        out.println("<count>" + p.getSaleCount() + "</count>");
        out.println("</product>");
}
out.println("</response>");
out.close();
}
}
```

数据库中商品的销售数量是不断变化的，本例实现的是每隔 5 秒查找数据库中销量前三名的商品信息，并在浏览器中进行显示。在测试运行过程中，我们可以通过直接在数据库中执行 update 语句修改商品的销量，以查看页面的刷新效果。

3.2　练习

1. 创建一张存储新闻标题的数据表，应包含字段：自增编号、新闻标题、发布日期时间。
2. 创建一个显示最新新闻的页面，该页面每隔 10 秒显示最新的前 5 条新闻的标题。

3.3 作业

完成一个简单的用户管理程序,使用 XMLHttpRequest 异步请求服务器完成如下功能:在用户列表页面上,当鼠标浮动在用户名上时则显示浮动窗口,显示用户的详细信息。

第4章 使用 Hibernate 完成对象持久化

本阶段目标

完成本章内容后,你将能够:
◇ 使用 Hibernate 完成持久化工作。

本阶段给出的步骤全面详细,请学员按照给出的上机步骤独立完成上机练习,以达到要求的学习目标。请认真完成下列步骤。

4.1 指导

本次上机的任务是使用 Hibernate 技术实现"我的宝贝儿"电子宠物网站的持久化层。训练的技能点主要包括:

(1) 学会使用 Hibernate 完成对象的持久化(给项目添加 Hibernate 支持,配置 Hibernate 配置文件,编写 Hibernate 映射文件和编写持久化代码)。

(2) 学会使用 MyEclipse 提供的工具简化 Hibernate 开发。

⊙ 小贴士

> 在"我的宝贝儿"电子宠物网站中,你可以给你领养的小宝贝起个温馨的或响亮的名字,可以对它倾诉,与它交流。同样精心地给予它关心、爱护。你可爱的小宝贝会让你的生活变得丰富多彩。它和我们现实中的宠物一样,需要每天喂食、洗澡、锻炼、学习,与你一起玩耍和被你抚摸。当失去你的关心时,它会伤心,生病,离家出走,甚至会死掉。这同时也在锻炼你的耐心和你的责任感。
>
> 电子宠物会记录下它的生日、岁数、体重、智力、力量、训练程度、饥饿度、清洁度、心情、零花钱等。
>
> 你可以进行以下几种训练活动:喂食、讲故事、游戏、学习、洗澡等。
>
> 电子宠物智力达到一定程度时会自己打工,为自己买衣服,甚至会升级出它自己的小保护神呢。
>
> 你还可以写宠物日记,记录你们俩的故事,写你向它倾诉的话。

本章使用的数据库仍然在 SqlServer 管理系统中创建,命名为"epet",其中包含宠物信息表（其表结构如图 4-1 所示）和宠物日记表两个表（其表结构如图 4-2 所示）。

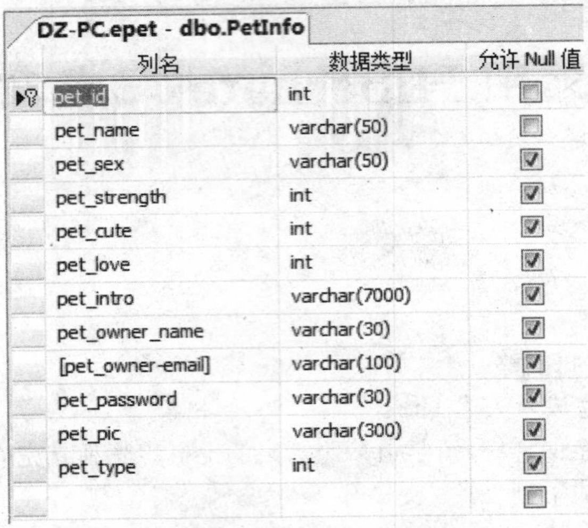

DZ-PC.epet - dbo.PetInfo

列名	数据类型	允许 Null 值
pet_id	int	☐
pet_name	varchar(50)	☐
pet_sex	varchar(50)	☑
pet_strength	int	☑
pet_cute	int	☑
pet_love	int	☑
pet_intro	varchar(7000)	☑
pet_owner_name	varchar(30)	☑
[pet_owner-email]	varchar(100)	☑
pet_password	varchar(30)	☑
pet_pic	varchar(300)	☑
pet_type	int	☑
		☐

图 4-1　宠物信息表

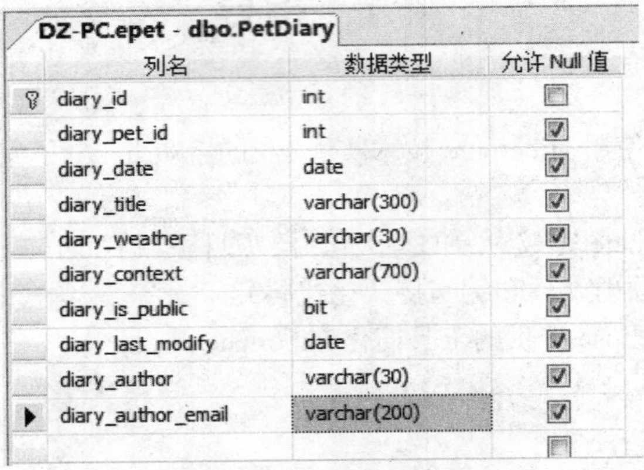

DZ-PC.epet - dbo.PetDiary

列名	数据类型	允许 Null 值
diary_id	int	☐
diary_pet_id	int	☑
diary_date	date	☑
diary_title	varchar(300)	☑
diary_weather	varchar(30)	☑
diary_context	varchar(700)	☑
diary_is_public	bit	☑
diary_last_modify	date	☑
diary_author	varchar(30)	☑
diary_author_email	varchar(200)	☑
		☐

图 4-2　宠物日记表

“我的宝贝儿”电子宠物网站首页的显示效果如图 4-3 所示。

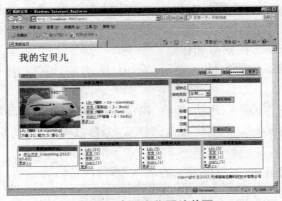

图 4-3　电子宠物网站首页

这个页面上囊括了"领养宠物""宠物登录""金宝宝排行""查找宠物""最新宠物日记"列表等业务模块。

点击"更多"链接,可以得到如图 4-4 所示的宠物列表页面。

图 4-4　宠物列表页面

在宠物列表页面中可以根据指定的条件搜索宠物信息,并查看宠物的总积分和各个属性值。

在首页上,点击左上角的"领养宠物"链接,可以得到如图 4-5 所示的页面。

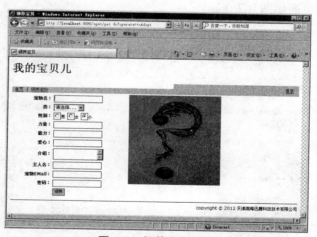

图 4-5　领养宠物页面

在领养宠物页面上输入领养信息,即可领养一个宝贝,并查看新宝贝信息。我们可以使用宝贝的编号和密码登录网站,给宝贝喂食、讲故事、让宝贝玩游戏增加聪明值,还可以进行日记编辑等工作。如图 4-6 所示。

294　　SSH 轻量级框架实践

图 4-6　宝贝明细页面

编写宠物日记的页面如图 4-7 所示。

图 4-7　编写宠物日记的页面

编辑完宠物日记点击"提交"后，该日记的标题会以链接的形式出现在宠物明细信息的页面上，如图 4-6 中的宠物日历所示。点击该链接可以看到如图 4-8 所示的日记详情页面。

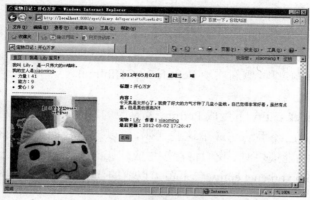

图 4-8　宠物日记详情页面

4.2　练习

分阶段完成以下任务：
● 阶段 1：使用 Hibernate 技术重新实现 PetInfo 相关的持久化代码。
训练要点：
（1）给项目添加 Hibernate 支持。
（2）学会修改 Hibernate 配置文件。
（3）编写 Hibernate 映射文件。
（4）使用 Hibernate 编写代码完成持久化工作。

根据本章理论部分的分析而知：我们可以使用 Hibernate 技术重新实现 PetInfo 相关的持久化代码。创建 PetInfoDAO 接口的 Hibernate 实现。实现思路及关键代码如下：

（1）使用 MyEclipse 工具给项目添加 Hibernate 支持。HibernateSessionFactory 放在 com.xtgj.epet.dao.hibimpl 包下。

（2）修改 Hibernate 配置文件，正确配置数据库连接，设置 show_sql 属性为 true。

（3）在 com.xtgj.epet.entity 包下创建 PetInfo.hbm.xml，完成 PetInfo 实体类到数据库 PetInfo 表的映射，并在 Hibernate 配置文件中添加相应 <mapping resource="…" /> 配置节点。

（4）创建 com.xtgj.epet.dao.hibimpl.PetInfoDAOHibImpl 类，实现 PetInfoDAO 接口。使用 Hibernate 技术实现除 search() 方法外的其他持久化方法。

（5）创建 com.xtgj.epet.dao.biz.PetInfoBizImpl 类，在 PetInfoBizImpl 类中，添加"private PetInfo DAOHibImpl petdao = new PetInfoDAOHibImpl();"。

（6）重新运行系统，验证领养宠物（pet_toAdopt.action）、查看宠物明细（pet_toView?petId=[id]）和编辑宠物信息（pet_toEdit.action?petId=[id]）功能是否正常运行。

● 阶段 2：使用 Hibernote 实现 PetDiary 相关持久化代码。
使用 Hibernate 技术重新实现 PetDiary 相关的持久化代码。创建 PetDiaryDAO 接口的 Hibernate 实现。

● 阶段3：简化Hibernate开发过程。

使用MyEclipse的Hibernate工具简化Hibernate开发过程。

使用BaseHibernateDAO简化DAO实现代码。

简化Hibernate开发过程，重新实现PetInfoDAOHibImpl。

实现思路及关键代码：

（1）在com.xtgj.epet. entity下生成PetInfo表对应的实体类和映射文件，阅读生成的代码。由于这个项目中，我们已经手动完成了这些代码，所以这次我们生成的代码并不应用到项目中。当需要开发新的功能模块时，我们将直接生成代码到com.xtgj.epet.entity包下。这步练习也可以直接生成到com.xtgj.epet.entity包下覆盖以前的代码。

注意：
　　由于下一章才学"关联"，为了不生成"关联"的相关代码，在如图4-9所示的对话框中，不要选最下面两个复选框。

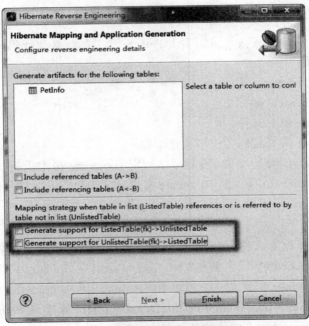

图4-9　最下面两个复选框不要选

（2）创建BaseHibernateDAO（代码参见理论部分第5章），简化DAO实现代码。

● 阶段4：简化PetDiaryDAOHibimpl代码。

重新生成PetDiary实体类和映射文件，使用BaseHibernateDAO简化PetDiaryDAOHibImpl代码。

下面给出示例代码4-1所示的部分代码。

示例代码 4-1：PetInfo.hbm.xml

```xml
<?xml version="1.0" encoding="utf-8"?>
<!DOCTYPE hibernate-mapping PUBLIC "-//Hibernate/Hibernate Mapping DTD 3.0//EN"
"http://www.hibernate.org/dtd/hibernate-mapping-3.0.dtd">
<!--
    Mapping file autogenerated by MyEclipse Persistence Tools
-->
<hibernate-mapping>
    <class name="com.xtgj.epet.entity.PetInfo" table="PetInfo" schema="dbo" catalog="epet">
        <id name="petId" type="java.lang.Integer">
            <coiumn name="pet_id" />
            <generator class="assigned" />
        </id>
        <property name="petName" type="java.lang.String">
            <column name="pet_name" length="50" not-null="true" />
        </property>
        <property name="petSex" type="java.lang.String">
            <column name="pet_sex" length="50" />
        </property>
        <property name="petStrength" type="java.lang.Integer">
            <column name="pet_strength" />
        </property>
        <property name="petCute" type="java.lang.Integer">
            <column name="pet_cute" />
        </property>
        <property name="petLove" type="java.lang.Integer">
            <column name="pet_love" />
        </property>
        <property name="petIntro" type="java.lang.String">
            <column name="pet_intro" length="7000" />
        </property>
        <property name="petOwnerName" type="java.lang.String">
            <column name="pet_owner_name" length="30" />
        </property>
        <property name="petOwnerEmail" type="java.lang.String">
            <column name=""pet_owner-email"" length="100" />
```

```xml
        </property>
        <property name="petPassword" type="java.lang.String">
            <column name="pet_password" length="30" />
        </property>
        <property name="petPic" type="java.lang.String">
            <column name="pet_pic" length="300" />
        </property>
        <property name="petType" type="java.lang.Integer">
            <column name="pet_type" />
        </property>
    </class>
</hibernate-mapping>
```

PetInfoDAOHibImpl 类如示例代码 4-2 所示。

示例代码 4-2：PetInfoDAOHibImpl 类

```java
package com.xtgj.epet.dao.hibimpl;

import java.util.List;
import com.xtgj.epet.dao.BaseHibernateDAO;
import com.xtgj.epet.dao.PetInfoDAO;
import com.xtgj.epet.entity.PetInfo;

public class PetInfoDAOHibImpl extends BaseHibernateDAO implements PetInfoDAO{
    public PetInfo load(int petId) {
        return (PetInfo) super.get(PetInfo.class, petId);
    }

    public void add(PetInfo PetInfo) {
        super.add(PetInfo);
    }

    public void del(int petId) {
        super.delete(PetInfo.class, petId);
    }

    public void update(PetInfo PetInfo) {
        super.update(PetInfo);
```

```java
        }

        @SuppressWarnings("unchecked")
        public List search(PetInfo condition) {
                List ret = super.search(PetInfo.class, condition);
                return ret;
        }

}
```

PetDiary 实体映射文件如示例代码 4-3 所示。

示例代码 4-3：PetDiary 实体映射文件代码

```xml
<?xml version="1.0" encoding="utf-8"?>
<!DOCTYPE hibernate-mapping PUBLIC "-//Hibernate/Hibernate Mapping DTD 3.0//EN"
"http://www.hibernate.org/dtd/hibernate-mapping-3.0.dtd">
<!-- 
    Mapping file autogenerated by MyEclipse Persistence Tools
-->
<hibernate-mapping>
    <class name="com.xtgj.epet.entity.PetDiary" table="PetDiary" schema="dbo" catalog="epet">
        <id name="diaryId" type="java.lang.Integer">
            <column name="diary_id" />
            <generator class="assigned" />
        </id>
        <property name="diaryPetId" type="java.lang.Integer">
            <column name="diary_pet_id" />
        </property>
        <property name="diaryDate" type="java.util.Date">
            <column name="diary_date" length="10" />
        </property>
        <property name="diaryTitle" type="java.lang.String">
            <column name="diary_title" length="300" />
        </property>
        <property name="diaryWeather" type="java.lang.String">
            <column name="diary_weather" length="30" />
        </property>
```

```xml
        <property name="diaryContext" type="java.lang.String">
            <column name="diary_context" length="700" />
        </property>
        <property name="diaryIsPublic" type="java.lang.Boolean">
            <column name="diary_is_public" />
        </property>
        <property name="diaryLastModify" type="java.util.Date">
            <column name="diary_last_modify" length="10" />
        </property>
        <property name="diaryAuthor" type="java.lang.String">
            <column name="diary_author" length="30" />
        </property>
        <property name="diaryAuthorEmail" type="java.lang.String">
            <column name="diary_author_email" length="200" />
        </property>
    </class>
</hibernate-mapping>
```

PetDiaryDAOHibImpl 类如示例代码 4-4 所示。

示例代码 4-4：PetDiaryDAOHibImpl 类代码

```java
package com.xtgj.epet.dao.hibimpl;

import java.util.List;
import com.xtgj.epet.dao.BaseHibernateDAO;
import com.xtgj.epet.dao.PetDiaryDAO;
import com.xtgj.epet.entity.PetDiary;

public class PetDiaryDAOHibImpl extends BaseHibernateDAO implements PetDiaryDAO{
    public PetDiary load(int diaryId) {
        return (PetDiary) super.get(PetDiary.class, diaryId);
    }

    public void add(PetDiary PetDiary) {
        super.add(PetDiary);
    }

    public void del(int diaryId) {
```

```java
        super.delete(PetDiary.class, diaryId);
    }

    public void update(PetDiary PetDiary) {
        super.update(PetDiary);
    }

    @SuppressWarnings("unchecked")
    public List search(PetDiary condition) {
        List ret = super.search(PetDiary.class, condition);
        return ret;
    }
}
```

PetInfoBizImpl 业务测试类如示例代码 4-5 所示。

示例代码 4-5：PetInfoBizImpl 类代码

```java
package com.xtgj.epet.dao.biz;

import com.xtgj.epet.dao.hibimpl.PetInfoDAOHibImpl;
import com.xtgj.epet.entity.PetInfo;

public class PetInfoBizImpl {
    private PetInfoDAOHibImpl petdao = new PetInfoDAOHibImpl();

    public static void main(String[] args) {
        PetInfoBizImpl imp = new PetInfoBizImpl();
        PetInfo pet1 = new PetInfo();
        pet1.setPetCute(1);
        pet1.setPetIntro("hello");
        pet1.setPetLove(1);
        pet1.setPetName("mary");
        pet1.setPetOwnerEmail("mary@163.com");
        pet1.setPetOwnerName("myName");
        pet1.setPetPassword("000000");
        pet1.setPetPic("images/pet/pig.jpg");
        pet1.setPetSex("m");
        pet1.setPetStrength(1);
```

```
            pet1.setPetType(1);
            imp.petdao.add(pet1);
    }
}
```

4.3 作业

实现简单的用户管理功能，要求如下：
（1）创建用户表（SysUser），含 3 个字段：username，password，flag。
（2）实现添加用户功能。
（3）使用 Hibernate 技术实现。
（4）使用 DispatchAction 和 Struts 标签。

提示：
- 使用 MyEclipse 生成实体类和映射文件，使用 BaseHibernateDAO 简化编码。
- 继续要求（1），实现用户编辑功能。
- 继续要求（2），显示已注册用户列表，并提供删除功能。

提示：
通过调用 BaseHibernateDAO 的 search() 方法可以得到全部用户的列表。代码如下：

```
List allUser=super.search(SysUser.class,new SysUser());
```

注意：
保存好完成的代码，后续章节的作业中将要求在上面完成代码的基础上扩充功能。

第 5 章　Hibernate 的关联映射

本阶段目标

完成本章内容后,你将能够:
◇ 掌握 Hibernate 映射关系。

本阶段给出的步骤全面详细,请学员按照给出的上机步骤独立完成上机练习,以达到要求的学习目标。请认真完成下列步骤。

5.1　指导

本次上机的任务是在上一章的基础上,继续开发"我的宝贝儿"电子宠物网站项目,要求如下:

(1)在 PetInfo 和 PetDiary 间配置一对多关系和多对一关系,并对应修改持久化代码,保证之前完成的功能正常运行。
(2)实现"根据日记 id 显示日记明细信息"功能。
(3)实现"删除宠物"的功能。
(4)简化宠物明细信息页面显示日记列表功能的实现。

训练的技能点包括以下两点:
(1)学会配置 Hibernate 一对多关系和多对一关系。
(2)学会在配置了一对多关系或多对一关系后如何编写持久化代码。

5.2　练习

分阶段完成以下任务:
● 阶段 1:添加 PetDiary 到 PetInfo 的多对一关联。
训练要点:
(1)配置 Hibernate 多对一关系。
(2)在配置了多对一关系的基础上完成持久化代码。

需求说明：

建立 PetDiary 到 PetInfo 的多对一关系，并对应修改持久化代码，保证之前完成的功能正常运行。添加多对一关联的步骤如下：

（1）在 PetDiary 实体类中去掉 petId 属性，增加"private PetInfo petInfo=null；"和相应的 getter 和 setter 方法，并修改因删掉 petId 属性引入的编译错误，现在将通过 petDiary.getPetInfo().getPetId() 的方式得到宠物日记对应的宠物编号。定义 AbstractPetDiary 类作为 PetDiary 实体类的抽象父类。如示例代码 5-1 所示。

示例代码 5-1：AbstractPetDiary.java 代码片段

```java
public abstract class AbstractPetDiary implements java.io.Serializable {
    private Integer diaryId;
    private PetInfo petInfo; // 添加多对一关联
    private Date diaryDate;
    private String diaryTitle;
    private String diaryWeather;
    private String diaryContext;
    private Byte diaryIsPublic;
    private Date diaryLastModify;
    private String diaryAuthor;
    private String diaryAuthorEmail;

    public AbstractPetDiary() {
    }

    public AbstractPetDiary(PetInfo petinfo, Date diaryDate, String diaryTitle,
            String diaryWeather, String diaryContext, Byte diaryIsPublic,
            Date diaryLastModify, String diaryAuthor, String diaryAuthorEmail) {
        this.petInfo = petinfo;
        this.diaryDate = diaryDate;
        this.diaryTitle = diaryTitle;
        this.diaryWeather = diaryWeather;
        this.diaryContext = diaryContext;
        this.diaryIsPublic = diaryIsPublic;
        this.diaryLastModify = diaryLastModify;
        this.diaryAuthor = diaryAuthor;
        this.diaryAuthorEmail = diaryAuthorEmail;
    }
```

```
    // 省略 getter/setter
}
```

PetDiary 如示例代码 5-2 所示。

示例代码 5-2：PetDiary.java 代码片段

```java
package com.xtgj.epet.entity;

import java.util.Date;
import com.xtgj.epet.comm.Util;

public class PetDiary extends AbstractPetDiary implements java.io.Serializable {
    private static final long serialVersionUID = 195817813043533293L;

    public PetDiary() {
        init();
    }

    public PetDiary(PetInfo petinfo, Date diaryDate, String diaryTitle,
            String diaryWeather, String diaryContext, Byte diaryIsPublic,
            Date diaryLastModify, String diaryAuthor, String diaryAuthorEmail) {
        super(petinfo, diaryDate, diaryTitle, diaryWeather, diaryContext,
                diaryIsPublic, diaryLastModify, diaryAuthor, diaryAuthorEmail);
        init();
    }

    private void init() {
        this.setDiaryLastModify(new Date());
        this.setDiaryDate(new Date());
        this.setDiaryIsPublic(new Byte((byte) 1));
    }

    // 以下为日期处理函数
    public String getDiaryDateString() {
        String ret = Util.formatDate(this.getDiaryDate());
        return ret;
    }
```

```java
public String getDiaryDateStringCN() {
    String ret = Util.formatDateCN(this.getDiaryDate());
    return ret;
}

public String getDiaryDateWeekName() {
    String ret = Util.getWeekNameCN(this.getDiaryDate());
    return ret;
}

public void setDiaryDateString(String diaryDate) {
    Date date = Util.parseDate(diaryDate);
    this.setDiaryDate(date);
}

public String getDiaryLastModifyString() {
    String ret = Util.formatTime(this.getDiaryLastModify());
    return ret;
}

public void setDiaryIsPublicString(String diaryIsPublicString) {
    Byte diaryIsPublic = null;
    if ("on".equals(diaryIsPublicString)) {
        diaryIsPublic = new Byte((byte) 1);
    } else {
        diaryIsPublic = new Byte((byte) 0);
    }
    super.setDiaryIsPublic(diaryIsPublic);
}

public String getDiaryIsPublicString() {
    Byte diaryIsPublic = super.getDiaryIsPublic();
    if (null != diaryIsPublic && diaryIsPublic.intValue() == 1) {
        return "on";
    } else {
        return "off";
    }
}
```

}

（2）修改 PetDiary.hbm.xml，去掉 petId 属性的配置，增加 <many-to-one> 节点配置。如示例代码 5-3 所示。

示例代码 5-3：PetDiary.hbm.xml 配置信息

```xml
<hibernate-mapping>
    <class name="com.xtgj.epet.entity.PetDiary" table="PetDiary" schema="dbo" catalog="epet">
        <id name="diaryId" type="java.lang.Integer">
            <column name="diary_id" />
            <generator class="native"></generator>
        </id>
        <many-to-one name="petInfo" class="PetInfo" fetch="select">
            <column name="diary_pet_id" />
        </many-to-one>
        <property name="diaryDate" type="java.util.Date">
            <column name="diary_date" length="0" />
        </property>
        <property name="diaryTitle" type="java.lang.String">
            <column name="diary_title" length="300" />
        </property>
        <property name="diaryWeather" type="java.lang.String">
            <column name="diary_weather" length="30" />
        </property>
        <property name="diaryContext" type="java.lang.String">
            <column name="diary_context" length="10000" />
        </property>
        <property name="diaryIsPublic" type="java.lang.Byte">
            <column name="diary_is_public" />
        </property>
        <property name="diaryLastModify" type="java.util.Date">
            <column name="diary_last_modify" length="0" />
        </property>
        <property name="diaryAuthor" type="java.lang.String">
            <column name="diary_author" length="30" />
        </property>
        <property name="diaryAuthorEmail" type="java.lang.String">
```

```xml
            <column name="diary_author_email" />
        </property>
    </class>

</hibernate-mapping>
```

(3)对应修改添加日记的 JSP 页面代码,将

```html
<input type="hidden" name="petId" value="${petId}" />
```

修改为

```html
<input type="hidden" name="petId" value="${petForm.item.petId}" />
```

重新运行添加宠物日记功能,验证修改成功。

(4)继承自 BaseDAO 重新实现 PetDiaryDAO 类,如示例代码 5-4 所示。

示例代码 5-4:PetDiaryDAO.java

```java
package com.xtgj.epet.dao;

// 省略 import

/**
 * 数据访问对象
 * @see .PetDiary
 * @author 迅腾国际
 */
public class PetDiaryDAO extends BaseDAO {
    public List list(String petId) {
        String hql = "from PetDiary d where d.petInfo.petId=" + petId;
        List ret = super.list(hql);
        return ret;
    }

    public void list(String listType, PetDiary condition, PageResult pageResult) {
        String hql = "from PetDiary d where 1=1 ";
        if (null != condition) {
            if (Util.isNotNullOrEmpty(condition.getDiaryTitle()))// 标题
            {
                hql += "and d.diaryTitle like '%" + condition.getDiaryTitle()
```

```java
                            + "%' ";
                }
                if (Util.isNotNullOrEmpty(condition.getDiaryAuthor()))// 作者
                {
                    hql += "and d.diaryAuthor like '%" + condition.getDiaryAuthor()
                            + "%' ";
                }
                if (Util.isNotNullOrEmpty(condition.getDiaryDateString()))// 日期
                {
                    hql += "and d.diaryDate = '" + condition.getDiaryDateString()
                            + "' ";
                }
                if (Util.isNotNullOrEmpty(condition.getDiaryContext()))// 关键字
                {
                    hql += "and d.diaryContext like '%"
                            + condition.getDiaryContext() + "%' ";
                }
            }
            if ("date".equals(listType) || null == listType) {
                hql += "order by d.diaryDate desc";
            } else if ("new".equals(listType)) {
                hql += "order by d.diaryLastModify desc";
            }
            super.list(hql, pageResult);
    }

    public PetDiary load(String id) {
            PetDiary ret = null;
            String hql = "from PetDiary d left join fetch d.petInfo where d.diaryId = "
                            + id;
            List list = super.list(hql);
            if (null != list && list.size() == 1) {
                    ret = (PetDiary) list.get(0);
            }
            return ret;
    }
```

```java
public void delete(String id) {
    PetDiary o = this.load(id);
    super.delete(o);
}

public List topList(String type, int count) {
    String hql = "from PetDiary d ";

    if ("all".equals(type)) {
        hql += "order by d.diaryLastModify desc";
    }
    Query query = this.getSession().createQuery(hql);
    query.setFirstResult(0);
    query.setMaxResults(count);
    List ret = query.list();
    return ret;
}

/*
 * 查出某个月份的日记,根据日期排序
 *
 * @param petId
 * @param month
 * @return
 */
public List list(String petId, int year, int month) {
    // 计算下一个月
    Calendar cal = Calendar.getInstance();
    cal.set(Calendar.YEAR, year);
    cal.set(Calendar.MONTH, month - 1);
    cal.add(Calendar.MONTH, 1);
    int nextYear = cal.get(Calendar.YEAR);
    int nextMonth = cal.get(Calendar.MONTH) + 1;
    // 构建 hql 语句
    String hql = "from PetDiary d where d.petInfo.petId=" + petId + " and "
            + "d.diaryDate >= '" + year + "-" + month + "-1' and "
            + "d.diaryDate < '" + nextYear + "-" + nextMonth + "-1'";
```

```
                    + " order by d.diaryDate asc";
            // 执行查询
            List ret = super.list(hql);
            // 返回结果
            return ret;
        }
    }
```

（5）DiaryBiz 业务类如示例代码 5-5 所示。

示例代码 5-5：DiaryBiz.java 源代码

```java
package com.xtgj.epet.biz;

// 省略 import

/*
 * Diary 业务逻辑类。
 * @author 迅腾国际
 */
public class DiaryBiz {
    private PetDiaryDAO diaryDAO = null;

    public PetDiaryDAO getDiaryDAO() {
        return diaryDAO;
    }

    public void setDiaryDAO(PetDiaryDAO diaryDAO) {
        this.diaryDAO = diaryDAO;
    }

    public List list(String petId) {
        List ret = this.getDiaryDAO().list(petId);
        return ret;
    }

    public void list(String petId, PetDiary condtion, PageResult pageResult) {
        this.getDiaryDAO().list(petId, condtion, pageResult);
    }
```

```java
public void doAdd(PetDiary item) {
    this.getDiaryDAO().save(item);
}

public void doUpdate(PetDiary item) {
    this.getDiaryDAO().update(item);
}

public void doDel(String id) {
    this.getDiaryDAO().delete(id);
}

public List topList(String type, int count) {
    List ret = this.getDiaryDAO().topList(type, count);
    return ret;
}

@SuppressWarnings("unchecked")
public HashMap list(String petId, int year, int month) {
    HashMap ret = new HashMap();
    List list = this.getDiaryDAO().list(petId, year, month);
    Iterator it = list.iterator();
    while (it.hasNext()) {
        PetDiary diary = (PetDiary) it.next();
        Calendar cal = Calendar.getInstance();
        cal.setTime(diary.getDiaryDate());
        String date = cal.get(Calendar.DAY_OF_MONTH) + "";
        if (ret.containsKey(date)) {
            List ls = (List) ret.get(date);
            ls.add(diary);
        } else {
            List ls = new ArrayList();
            ls.add(diary);
            ret.put(date, ls);
        }
    }
    return ret;
}
```

```
public PetDiary load(String id) {
    PetDiary ret = this.getDiaryDAO().load(id);
    return ret;
}
```

- 阶段 2：实现显示日记详细信息的功能。

需求说明：

根据日记 ID 显示一篇日记的详细信息（diary_toView?petId=[petId]）。需要显示宠物的名称，并显示为链接，链接到宠物详细信息页面。显示宠物的图片如图 5-1 所示。

图 5-1　宠物详细信息

提示：

（1）新建 /WEB-INF/jsp/diary/detail.jsp，如示例代码 5-6 所示。

示例代码 5-6：detail.jsp 代码片段

```html
<div id="diaryDiv">
    <p>
            <br />
            <br />
            <b> ${diaryForm.item.diaryDateStringCN}    
                ${diaryForm.item.diaryDateWeekName}    
```

```
                    ${diaryForm.item.diaryWeather} </b>
        </p>
        <p>
            <b>标题:</b>${diaryForm.item.diaryTitle}
        </p>
        <p>
            <b>内容:</b>
        <div>
            ${diaryForm.item.diaryContext}
        </div>

        <p>
            <b>宠物:</b>
<a href="pet/pet_toBaby?petId=${diaryForm.item.petInfo.petId}">
${diaryForm.item.petInfo.petName}</a>  
            <b>作者:</b>
<a href="mailto:${diaryForm.item.diaryAuthorEmail}">${diaryForm.item.diaryAuthor}</a>
            <br />
            <b>最后更新:</b>${diaryForm.item.diaryLastModifyString}
            <br />
        </p>
        <button onclick="javascript:history.go(-1);">
            返回
        </button>
</div>
```

（2）在 struts.xml 配置文件中 diary 包下，增加到 list 页面的转发配置如示例代码 5-7 所示。

示例代码 5-7：struts.xml
```
<package name="diary" namespace="/diary" extends="struts-default">
    <action name="diary_*" class="diary" method="{1}">
        <result name="list">/WEB-INF/jsp/diary/list.jsp
        </result>
    </action>
</package>
```

（3）在 DiaryAction 中增加 toView 方法，如示例代码 5-8 所示。

示例代码 5-8：DiaryAction.java

```java
public String toView() {
    PetDiary petDiary = this.getDiaryBiz().load(diaryForm.getId());
    diaryForm.setItem(petDiary);
    PetInfo petInfo = petDiary.getPetInfo();
    PetForm petForm = new PetForm();
    petForm.setItem(petInfo);
    ActionContext.getContext().put("petForm", petForm);
    return "detail";
}
```

- 阶段 3：添加 PetInfo 到 PetDiary 的一对多关联。

需求说明：

1. 配置一对多关联。
2. 实现删除宠物的功能。

实现思路及关键代码：

（1）配置一对多关系

首先，在 PetInfo 实体类中增加属性：

```java
private List petDiaries = new ArrayList();
```

与理论课介绍的内容稍稍不同，类型是 List 而不是 Set。在对应的 PetInfo.hbm.xml 映射文件中，增加如下配置节点，如示例代码 5-9 所示。

示例代码 5-9：配置节点

```xml
<bag name="petDiaries" inverse="true">
    <key>
        <column name="diary_pet_id" />
    </key>
    <one-to-many class="PetDiary" />
</bag>
```

`<bag>` 节点而不是理论课介绍的 `<set>` 节点。两种方式都可以，如果更习惯在程序中使用 List 可采用这种方式。

（2）实现删除宠物功能

DAO 层中增加 public void del(Long id) 方法；Biz 层增加 public void del(Long id) 方法；PetAction 中增加 doDel() 方法。删除成功后重定向到首页。

参考解决方案：

DAO 中 doDel() 方法的实现如示例代码 5-10 所示。

> 示例代码 5-10：doDel 方法源代码片段
>
> ```
> public void doDel(Long id){
>     super.del(PetInfo.class,id);
> }
> ```

由于关联的 cascade 的属性设置为 all，Hibernate 会自动删除级联的日记数据。

● 阶段 4：简化宠物详细信息页显示日记列表功能的实现。

需求说明：

由于配置了一对多关联，加载宠物数据时已经加载了关联的日记数据，无需再查一次。要求修改 PetAction 中 toView 代码和页面相关代码，将多余的查询操作去掉，效果不变。如图 5-2 所示。

图 5-2　日记数据

5.3　作业

1. 某公司的办公系统中有两个数据库表：职员表 (emp_id, emp_name)，奖励表 (pri_id, pri_emp_id, pri_year, pri_type, pri_comment)。奖励类别包括年度最佳质量奖、年度最佳新人、年度最佳员工、年度最佳经理。该系统要求实现如下功能：

显示职员列表，要求显示职员编号、职员姓名、荣获奖励。格式如表 5-1 所示。

表 5-1　职员列表

职员编号	职员姓名	荣获奖励
1	刘合平	年度最佳质量奖（2005） 年度最佳员工（2006） 年度最佳经理（2007）

续表

职员编号	职员姓名	荣获奖励
2	周扬	年度最佳新人（2006） 年度最佳质量奖（2007）
3	马旭	年度最佳新人（2007）

如何使用 Hibernate 技术实现这个功能呢？

2. 在第 1 题描述的系统中，实现如下功能：显示当年所有人的获奖记录。格式如表 5-2 所示：

表 5-2　获奖信息表

编号	年度	类型	获奖员工	评语
26	2007	年度最佳质量奖	马旭	带领测试团队，保证了 x 项目和 xx 项目的稳步进行和顺利验收，而且还创造了 x 系统无脱机运行超万小时的新纪录
27	2007	年度最佳新人	刘海超	勤奋学习，爱岗敬业，技术水平高超。在极为不利的情况下，顽强奋战了两个多月，使 x 项目得以验收。其技术水平，团队表现和高度负责的精神超出了预期
28	2007	年度最佳员工	周扬	长期出差在库尔勒油田，不怕苦，不怕累，客户关系融洽，行业积累深厚，对 xx 项目中标发挥了很大作用
29	2007	年度最佳经理	刘合平	全面参与实施了 x 项目，xx 项目，并参加了 x 项目、xx 项目的售前支持工作，其项目控制能力和团队管理能力堪称典范。特别是 x 项目，项目成员以新人为主，在顺利完成项目的同时帮助新人迅速成长，为部门和公司创造了巨大的价值

提示：配置多对一，可以方便地显示获奖员工姓名。

第 6 章　Hibernate 查询

本阶段目标

完成本章内容后,你将能够:
- 掌握 Hibernate HQL 查询技术。
- 掌握 Hibernate Criteria 查询技术。

本阶段给出的步骤全面详细,请学员按照给出的上机步骤独立完成上机练习,以达到要求的学习目标。请认真完成下列步骤。

6.1　指导

本次上机的任务是使用 Hibernate HQL 技术和 Criteria 查询技术实现宠物查询。
训练的技能点:
(1) HQL 查询。
(2) Criteria 查询。

6.2　练习

分阶段完成以下任务:
● 阶段 1:采用 HQL 方式实现根据宠物名称查询宠物,查询结果按指定方式排序。
训练要点:
HQL 查询。
需求说明:
(1) 实现根据宠物名称查询的功能:不输入查询条件则返回所有记录,输入查询条件则按照宠物名称模糊查询。
(2) 用户以下拉列表框的方式指定排序方式(按活力、聪明、可爱,"活力"对应"力量","可爱"对应"爱心",默认则按宠物名称排序),提交查询后返回结果按照指定方式排序。如图 6-1 所示。

第6章 Hibernate 查询

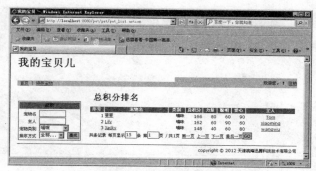

图 6-1 宠物列表查询

实现思路及关键代码：

在原有宠物列表的功能（第 5 章上机已完成）上完成按名称模糊查询功能：

（1）在 PetForm 中增加类型为 com.xtgj.epet.entity.PetInfo 的 condition 属性，用于接收页面传入的查询条件。

（2）在页面（/WEB-INF/jsp/pet/list.jsp）增加标签和输入框。

（3）修改 com.xtgj.epet.dao.PetInfoDAO 类中的 list 方法。

实现排序功能：

（1）在 PetForm 中增加 String 类型的 listType 属性。

（2）在页面上增加下拉列表框代码。

（3）修改对应 Biz 和 Dao 的 list 方法，增加一个 listType 参数。

参考解决方案：

PetForm 中增加的代码如示例代码 6-1 所示。

示例代码 6-1：PetForm 代码片段

```
package com.xtgj.epet.web.form;

public class PetForm extends BaseForm {
    private static final long serialVersionUID = -2302826642547539235L;

    private PetInfo item = null;
    private PetInfo condition = null;// 增加 condition 属性
    private String listType = "all";// 增加 listType 属性
    private String trainingType = "";

    public PetForm() {
        if (null == this.item) {
            this.item = new PetInfo();
        }
        if (null == this.condition) {
```

```
                    this.condition = new PetInfo();
            }
    }

    // 省略 getter/setter

}
```

宠物列表页面 list.jsp 如示例代码 6-2 所示。

示例代码 6-2：list.jsp 代码

```
<body>
<s:form action="pet/pet_list.action">
    <input type="hidden" name="listType" value="<%=request.getParameter("listType") %>" />
    <jsp:include flush="true" page="/header.jsp"></jsp:include>
    <div class="main_div">
    <table>
            <tr>
                <td width="200px" valign="top">
                <br />
                <br />
                <br />
                <table style="border:solid 1px black;">
                <tr>
                        <th style="background-color:gray;height:20px;">
                        查找
                        </th>
                </tr>
                <tr>
                    <td>
                                宠物名  
                            <s:textfield name="petForm.condition.petName" size="10"></s:textfield>
                            <br />
                                 
                            主人  
                             <s:textfield name="petForm.condition.petOwnerName" size="10"></s:textfield>
```

```
                                  <br />
                                  宠物类别  
                                  <s:select list="#{'-1':' 请选择 ...','1':' 千禧猪 ','2':' 喵咪 ','3':' 哥斯拉 '}"
name="petForm.condition.petType" listKey="key"
listValue="value">
                                                                            < /
s:select>
                                  <br />
                                  排序方式  
                                  <s:select list="#{'all':' 全 部 ...','strength':' 活 力 ','cute':' 聪明 ','love':' 可爱 '}"
name="petForm.listType" listKey="key"
listValue="value">
                                                                            < /
s:select>
                                          <button onclick="javascript:search-
Pet();"> 查找 </button>
                                  </td>
                       </tr>
                       </table>
                  </td>
                  <td>
                  <div>
                       <div class="input_title"><%
                            String listType = request.getParameter("petForm.
listType");
                            if ("all".equals(listType))
                            {
                                 out.print(" 总积分排名 ");
                            }else if ("cute".equals(listType))
                            {
```

```
                                out.print(" 最聪明排名 ");
                        }else if ("love".equals(listType))
                        {
                                out.print(" 最可爱排名 ");
                        }else if ("strength".equals(listType))
                        {
                                out.print(" 最活力排名 ");
                        }else
                        {
                                out.print(" 可爱的宝贝们！");
                        }
            %></div>


            <%
                        Integer oIndex = (Integer)request.getAttribute("petForm.pageResult.firstRec");
                        int index = oIndex.intValue();
            %>
            <table class="data_table">
            <tr>
                        <th width="40px"> 序号 </th>
                        <th width="200px"> 宠物名 </th>
                        <th width="50px"> 类别 </th>
                        <th width="50px"> 总积分 </th>
                        <th width="40px"> 力量 </th>
                        <th width="40px"> 聪明 </th>
                        <th width="40px"> 爱心 </th>
                        <th width="150px"> 主人 </th>
            </tr>

            <s:iterator var="item" value="petForm.pageResult.list">
                        <tr>
                                <td style="text-align:right;">
                                        <%=++index %>
                                </td>
                                <td>
```

```
                                                <a href="pet.do?operate=toBaby&id=${item.petId}" target="_blank">
                                                    ${item.petName}
                                                </a>
                                            </td>
                                            <td style="text-align:right;">${item.petTypeString}</td>
                                            <td style="text-align:right;">${item.petScore}</td>
                                            <td style="text-align:right;">${item.petStrength}</td>
                                            <td style="text-align:right;">${item.petCute}</td>
                                            <td style="text-align:right;">${item.petLove}</td>
                                            <td style="text-align:center;"><a href="mailto:${item.petOwnerEmail}">${item.petOwnerName}</a></td>
                                        </tr>
                                    </s:iterator>

                                </table>
                                共 ${diaryForm.pageResult.recTotal} 条记录
                                每页显示 <input name="petForm.pageResult.pageSize" value="${petForm.pageResult.pageSize}" size="3" /> 条
                                第 <input name="petForm.pageResult.pageNo" value="${petForm.pageResult.pageNo}" size="3" /> 页
                                / 共 ${petForm.pageResult.pageTotal} 页
                                <a href="javascript:page_first();"> 第一页 </a>
                                <a href="javascript:page_pre();"> 上一页 </a>
                                <a href="javascript:page_next();"> 下一页 </a>
                                <a href="javascript:page_last();"> 最后一页 </a>
                                <button onclick="javascript:page_go();">GO</button>

                                <script>
                                    function page_go()
                                    {
                                        page_validate();
                                        document.forms[0].submit();
```

```js
            }
            function page_first()
            {
                document.forms[0].elements["petForm.pageResult.pageNo"].value = 1;
                document.forms[0].submit();
            }
            function page_pre()
            {
                var pageNo = document.forms[0].elements["petForm.pageResult.pageNo"].value;
                document.forms[0].elements["petForm.pageResult.pageNo"].value = parseInt(pageNo) - 1;
                page_validate();
                document.forms[0].submit();
            }
            function page_next()
            {
                var pageNo = document.forms[0].elements["petForm.pageResult.pageNo"].value;
                document.forms[0].elements["petForm.pageResult.pageNo"].value = parseInt(pageNo) + 1;
                page_validate();
                document.forms[0].submit();
            }
            function page_last()
            {
                document.forms[0].elements["petForm.pageResult.pageNo"].value = ${petForm.pageResult.pageTotal};
                document.forms[0].submit();
            }
            function page_validate()
            {
                var pageTotal = ${petForm.pageResult.pageTotal};
                var pageNo = document.forms[0].elements["petForm.pageResult.pageNo"].value;
                if (pageNo<1)pageNo=1;
                if (pageNo>pageTotal)pageNo=pageTotal;
```

```
                        document.forms[0].elements["petForm.pageResult.pageNo"].value
= pageNo;

                        var recTotal = ${petForm.pageResult.recTotal};
                        var pageSize = document.forms[0].elements["petForm.pageResult.
pageSize"].value;

                        if (pageSize<1)pageSize=1;
                        document.forms[0].elements["petForm.pageResult.pageSize"].val-
ue = pageSize;
                   }

            </script></div>
                        </td>
                   </tr>
            </table>
            </div>
            <jsp:include flush="true" page="/footer.jsp"></jsp:include>
      </s:form>
      </body>
```

- 阶段 2：采用 HQL 方式实现根据主人名和宠物类别查询宠物。

需求说明：

（1）根据宠物主人名模糊查询。

（2）根据宠物类别精确查询。如图 6-2 所示。

图 6-2 查找宠物信息

提示：

（1）修改页面。增加查询条件输入框中下拉列表框。

（2）修改对应 DAO 实现类 list 方法。增加对查询条件的支持。

（3）宠物类别查询条件的值为 -1 表示查询全部。

- 阶段 3：采用 Criteria 方式实现根据宠物名称查询宠物，查询结果按指定方式排序。

训练要点:Criteria 查询。

实现思路及关键代码:

只需要修改 PetInfoDAO 中的 list 方法如示例代码 6-3 所示。

示例代码 6-3:PetInfoDAO 中的采用 HQL 实现的 list 方法代码片段

```java
public void list(String listType,PetInfo condition,PageResult pageResult){
    String hql = "from PetInfo p where 1=1 ";
    if (null!=condition)
    {
        if (Util.isNotNullOrEmpty(condition.getPetName()))
        {
            hql += "and p.petName like '%" + condition.getPetName() + "%' ";
        }
        if (Util.isNotNullOrEmpty(condition.getPetType()))
        {
            hql += "and p.petType = " + condition.getPetType() + " ";
        }
        if (Util.isNotNullOrEmpty(condition.getPetOwnerName()))
        {
            hql += "and p.petOwnerName like '%" + condition.getPetOwnerName() +"%' ";
        }
    }
    if ("all".equals(listType) || null==listType)
    {
        hql += "order by p.petScore desc";
    }else if ("cute".equals(listType))
    {
        hql += "order by p.petCute desc";
    }else if ("love".equals(listType))
    {
        hql += "order by p.petLove desc";
    }else if ("strength".equals(listType))
    {
        hql += "order by p.petStrength desc";
    }
    super.list(hql, pageResult);
}
```

PetInfoDAO 中的 topList 方法,如示例代码 6-4 所示。

示例代码 6-4: PetInfoDAO 中的 topList 方法代码片段

```java
public List topList(String listType, int count) {
    String hql = "from PetInfo p ";

    if ("all".equals(listType))
    {
        hql += "order by p.petScore desc";
    }else if ("cute".equals(listType))
    {
        hql += "order by p.petCute desc";
    }else if ("love".equals(listType))
    {
        hql += "order by p.petLove desc";
    }else if ("strength".equals(listType))
    {
        hql += "order by p.petStrength desc";
    }
    Query query = this.getSession().createQuery(hql);
    query.setFirstResult(0);
    query.setMaxResults(count);
    List ret = query.list();
    return ret;
}
```

● 阶段 4:采用 Criteria 方式实现根据主人名和宠物类别查询宠物。
需求说明:
同阶段 2,采用 Criteria 查询实现,如示例代码 6-5 所示。

示例代码 6-5: PetInfoDAO 中的采用 Criteria 技术实现 list 方法代码片段

```java
if(condtion.getPetType()!=null&&condition.getPetType()!=-1){
  c.add(Restrictions.eq("petType",condition.getPetType()));
……
}
```

6.3 作业

1. 继续第 5 章的作业题,在已注册用户列表功能基础上实现如下功能:
（1）按照用户名模糊查询。
（2）根据 flag 进行精确查询（flag 为 -1,则用户状态为"已禁用";为 0,用户状态为"正常";为 1,用户状态为"VIP 用户"）。默认显示状态为"正常"和"VIP"的记录。
（3）使用 HQL 技术实现。
2. 使用 Criteria 查询实现上面的需求。
3. 在完成第 2 题的结果上增加排序功能:用户单击列表标题则按照该标题升序排序。

第 7 章 Hibernate 高级特性

本阶段目标

完成本章内容后,你将能够:
- ◇ 掌握 Hibernate 延迟加载。
- ◇ 掌握 Hibernate 缓存配置方式。

本阶段给出的步骤全面详细,请学员按照给出的上机步骤独立完成上机练习,以达到要求的学习目标。请认真完成下列步骤。

7.1 指导

本次上级任务是在上一章完成任务的基础上,继续开发"我的宝贝儿"电子宠物网站项目,要求如下:
(1)使用 Hibernate 延迟加载机制实现属性延迟加载。
(2)通过添加二级缓存插件,使得宠物信息对象和宠物日记对象能存放于二级缓存中。
(3)实现领养宠物功能。

训练的技能点:
(1)Hibernate 延迟加载机制。
(2)Hibernate 缓存机制。

7.2 练习

分阶段完成以下任务:
● 阶段 1:采用 Hibernate 延迟加载机制实现宠物信息延迟加载。
训练要点:
Hibernate 延迟加载机制。
需求说明:
实现需要宠物信息时才执行数据加载操作。

参考解决方案：

PetInfo.hbm.xml 配置文件如示例代码 7-1 所示。

示例代码 7-1：PetInfo.hbm.xml

```xml
<hibernate-mapping>
    <class name="com.xtgj.epet.entity.PetDiary" table="PetDiary" schema="dbo" catalog="epet">
        <id name="diaryId" type="java.lang.Integer">
            <column name="diary_id" />
            <generator class="native"></generator>

        </id>
        <many-to-one name="petInfo" class="PetInfo" fetch="select">
            <column name="diary_pet_id" />
        </many-to-one>
        <property name="diaryDate" type="java.util.Date">
            <column name="diary_date" length="0" />
        </property>
        <property name="diaryTitle" type="java.lang.String">
            <column name="diary_title" length="300" />
        </property>
        <property name="diaryWeather" type="java.lang.String">
            <column name="diary_weather" length="30" />
        </property>
        <property name="diaryContext" type="java.lang.String">
            <column name="diary_context" length="10000" />
        </property>
        <property name="diaryIsPublic" type="java.lang.Byte">
            <column name="diary_is_public" />
        </property>
        <property name="diaryLastModify" type="java.util.Date">
            <column name="diary_last_modify" length="0" />
        </property>
        <property name="diaryAuthor" type="java.lang.String">
            <column name="diary_author" length="30" />
        </property>
        <property name="diaryAuthorEmail" type="java.lang.String">
            <column name="diary_author_email" />
        </property>
```

```
        </class>
    </hibernate-mapping>
```

● 阶段2：采用Hibernate延迟加载机制实现宠物日记信息延迟加载。

训练要点：

Hibernate延迟加载机制。

需求说明：

实现需要宠物日记信息时才执行数据加载操作。

参考解决方案：

PetDiary.hbm.xml配置文件如示例代码7-2所示。

示例代码7-2：PetDiary.hbm.xml配置文件

```xml
<?xml version="1.0" encoding="utf-8"?>
<!DOCTYPE hibernate-mapping PUBLIC "-//Hibernate/Hibernate Mapping DTD 3.0//EN"
"http://www.hibernate.org/dtd/hibernate-mapping-3.0.dtd">
<!--
    Mapping file autogenerated by MyEclipse Persistence Tools
-->
<hibernate-mapping>
    <class name="com.xtgj.epet.entity.PetDiary" table="PetDiary" schema="dbo" catalog="epet" lazy="true">
        <id name="diaryId" type="java.lang.Integer">
            <column name="diary_id" />
            <generator class="assigned" />
        </id>
        <property name="diaryPetId" type="java.lang.Integer">
            <column name="diary_pet_id" />
        </property>
        <property name="diaryDate" type="java.util.Date">
            <column name="diary_date" length="10" />
        </property>
        <property name="diaryTitle" type="java.lang.String">
            <column name="diary_title" length="300" />
        </property>
        <property name="diaryWeather" type="java.lang.String">
            <column name="diary_weather" length="30" />
        </property>
```

```xml
        <property name="diaryContext" type="java.lang.String">
            <column name="diary_context" length="700" />
        </property>
        <property name="diaryIsPublic" type="java.lang.Boolean">
            <column name="diary_is_public" />
        </property>
        <property name="diaryLastModify" type="java.util.Date">
            <column name="diary_last_modify" length="10" />
        </property>
        <property name="diaryAuthor" type="java.lang.String">
            <column name="diary_author" length="30" />
        </property>
        <property name="diaryAuthorEmail" type="java.lang.String">
            <column name="diary_author_email" length="200" />
        </property>
    </class>
</hibernate-mapping>
```

● 阶段3：添加二级缓存插件，使得宠物信息对象能存放于二级缓存中。

训练要点：

Hibernate 二级缓存。

解决方案：

修改 hibernate.cfg.xml 配置文件，如示例代码 7-3 所示。

示例代码 7-3：hibernate.cfg.xml 配置文件源代码片段

```xml
<hibernate-configuration>
    <session-factory>
        <property name="connection.password">123456</property>
        <property name="connection.username">sa</property>
        <property name="connection.url">jdbc:sqlserver://127.0.0.1:1433;databaseName=advance</property>
        <property name="dialect">org.hibernate.dialect.SQLServerDialect</property>
        <property name="myeclipse.connection.profile">db1</property>
        <property name="connection.driver_class">com.microsoft.sqlserver.jdbc.SQLServerDriver</property>
        <property name="show_sql">true</property>
        <property name="hibernate.cache.provider_class">
            org.hibernate.cache.EhCacheProvider
```

```xml
        </property>
        <!-- 启动"查询缓存" -->
        <property name="hibernate.cache.use_query_cache">
            true
        </property>
        <!-- 省略部分代码 -->
    </session-factory>
</hibernate-configuration>
```

新建缓存配置文件,命名为 ehcache.xml,如示例代码 7-4 所示。

示例代码 7-4:ehcache.xml 配置文件源代码片段

```xml
<ehcache>
    <!--
    maxElementsInMemory 为缓存对象的最大数目,
    eternal 设置是否永远不过期,timeToIdleSeconds 对象处于空闲状态的最多秒数,
    timeToLiveSeconds 对象处于缓存状态的最多秒数
    -->
    <diskStore path="java.io.tmpdir"/>
    <defaultCache maxElementsInMemory="10000" eternal="false" timeToIdleSeconds="300"
    timeToLiveSeconds="600" overflowToDisk="true"/>
</ehcache>
```

在 PetInfo.hbm.xml 配置文件中配置缓存信息,如示例代码 7-5 所示。

示例代码 7-5:PetInfo.hbm.xml

```xml
<?xml version="1.0" encoding="utf-8"?>
<!DOCTYPE hibernate-mapping PUBLIC "-//Hibernate/Hibernate Mapping DTD 3.0//EN"
"http://www.hibernate.org/dtd/hibernate-mapping-3.0.dtd">
<!--
    Mapping file autogenerated by MyEclipse Persistence Tools
-->
<hibernate-mapping>
    <class name="com.xtgj.epet.entity.PetInfo" table="PetInfo" schema="dbo" catalog="epet" lazy="true">
        <cache usage="read-write"/>
        <id name="petId" type="java.lang.Integer">
            <column name="pet_id" />
```

```xml
            <generator class="assigned" />
        </id>
        <property name="petName" type="java.lang.String">
            <column name="pet_name" length="50" not-null="true" />
        </property>
        <property name="petSex" type="java.lang.String">
            <column name="pet_sex" length="50" />
        </property>
        <property name="petStrength" type="java.lang.Integer">
            <column name="pet_strength" />
        </property>
        <property name="petCute" type="java.lang.Integer">
            <column name="pet_cute" />
        </property>
        <property name="petLove" type="java.lang.Integer">
            <column name="pet_love" />
        </property>
        <property name="petIntro" type="java.lang.String">
            <column name="pet_intro" length="7000" />
        </property>
        <property name="petOwnerName" type="java.lang.String">
            <column name="pet_owner_name" length="30" />
        </property>
        <property name="petOwnerEmail" type="java.lang.String">
            <column name="`pet_owner-email`" length="100" />
        </property>
        <property name="petPassword" type="java.lang.String">
            <column name="pet_password" length="30" />
        </property>
        <property name="petPic" type="java.lang.String">
            <column name="pet_pic" length="300" />
        </property>
        <property name="petType" type="java.lang.Integer">
            <column name="pet_type" />
        </property>
    </class>
</hibernate-mapping>
```

- 阶段4：添加二级缓存插件，使得宠物日记信息能存放于二级缓存中。

训练要点:

Hibernate 二级缓存。

实现思路及关键代码:

在阶段 3 配置文件的基础上,修改 PetInfo.hbm.xml 配置文件中配置缓存信息,如示例代码 7-6 所示。

示例代码 7-6:PetDiary 实体映射文件代码

```xml
<?xml version="1.0" encoding="utf-8"?>
<!DOCTYPE hibernate-mapping PUBLIC "-//Hibernate/Hibernate Mapping DTD 3.0//EN"
"http://www.hibernate.org/dtd/hibernate-mapping-3.0.dtd">
<!--
    Mapping file autogenerated by MyEclipse Persistence Tools
-->
<hibernate-mapping>
    <class name="com.xtgj.epet.entity.PetDiary" table="PetDiary" schema="dbo" catalog="epet">
        <cache usage="read-write"/>
        <id name="diaryId" type="java.lang.Integer">
            <column name="diary_id" />
            <generator class="assigned" />
        </id>
        <property name="diaryPetId" type="java.lang.Integer">
            <column name="diary_pet_id" />
        </property>
        <property name="diaryDate" type="java.util.Date">
            <column name="diary_date" length="10" />
        </property>
        <property name="diaryTitle" type="java.lang.String">
            <column name="diary_title" length="300" />
        </property>
        <property name="diaryWeather" type="java.lang.String">
            <column name="diary_weather" length="30" />
        </property>
        <property name="diaryContext" type="java.lang.String">
            <column name="diary_context" length="700" />
        </property>
        <property name="diaryIsPublic" type="java.lang.Boolean">
            <column name="diary_is_public" />
```

```xml
        </property>
        <property name="diaryLastModify" type="java.util.Date">
            <column name="diary_last_modify" length="10" />
        </property>
        <property name="diaryAuthor" type="java.lang.String">
            <column name="diary_author" length="30" />
        </property>
        <property name="diaryAuthorEmail" type="java.lang.String">
            <column name="diary_author_email" length="200" />
        </property>
    </class>
</hibernate-mapping>
```

● 阶段5：实现领养宠物功能。

需求说明：

在领养宠物页面上输入领养信息，即可领养一个宝贝并显示宝贝信息。领养宠物页面如图7-1所示。

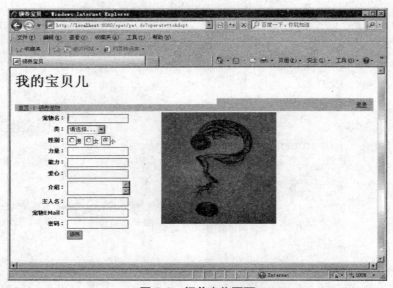

图 7-1 领养宠物页面

提示：

（1）在 /WEB-INF/jsp/ pet 中新建 petadoption.jsp 文件，如示例代码 7-7 所示。

示例代码 7-7：petadoption.jsp 源代码

```html
<body>
<div class="main_div">
    <table>
        <s:form action="pet/pet_adoption.action">
            <tr>
                <td width="200px" valign="top">
                <br />
                <br />
                <br />
                <table style="border:solid 1px black;">
                    <tr>
                        <td>
                                宠物名  
                            <s:textfield name="petForm.condition.petName" size="10"></s:textfield>
                            <br />
                                类  
                            <s:select list="#{'-1':' 请选择 ...','1':' 千禧猪 ','2':' 喵咪 ','3':' 哥斯拉 '}" name="petForm.condition.petType" listKey="key" listValue="value">
                            </s:select>
                            <br />
                                性别： 
                            <s:radio list="{' 男 ',' 女 ',' 小 '}" name=" petSex " />
                                力量： 
                            <s:textfield name="petForm.condition.petOwnerPower" size="10"></s:textfield>
                            <br />
                                能力： 
                            <s:textfield name="petForm.condition. petStrength " size="10"></s:textfield>
                            <br />
                                爱心： 
                            <s:textfield name="petForm.condition. petLove " size="10"></s:textfield>
                            <br />
                                介绍： 
                            <s:textfield name="petForm.condition. petIntro " size="10"></s:textfield>
                            <br />
                                主人名： 
                            <s:textfield name="petForm.condition.petOwnerName" size="10"></s:textfield>
```

```
                              <br />
                        宠物 EMail： 
            <s:textfield name="petForm.condition. petOwnerEmail " size="10"></s:textfield>
                              <br />
                        密码： 
            <s:textfield name="petForm.condition. petPassword " size="10"></s:textfield>
                              <br />

                     <button onclick="javascript:searchPet();"> 领养 </button>
                        </td>
                    </tr>
                </table>
<jsp:include flush="true" page="/footer.jsp"></jsp:include>
</s:form>
</body>
```

（2）在 struts.xml 配置文件中 pet 包下，增加 detail 页面的转发配置，如示例代码 7-8 所示。

示例代码 7-8：struts.xml 代码片段

```
<package name="pet" namespace="/pet" extends="struts-default">
        <action name="pet_*" class="pet" method="{1}">
            <result name="list">/WEB-INF/jsp/pet/detail.jsp
            </result>
        </action>
</package>
```

（3）在 PetAction 中增加 addPet( ) 方法，如示例代码 7-9 所示。

示例代码 7-9：PetAction.java 片段代码

```
public String addPet() {
        PetInfo pet = this.getPetBiz().load(petForm.getId());
        petForm.setItem(Pet);
        PetInfo petInfo = petForm.getPetInfo();
        PetForm petForm = new PetForm();
        petForm.setItem(petInfo);
        ActionContext.getContext().put("petForm", petForm);
        return "pet";
}
```

7.3 作业

继续第 6 章的作业题,实现如下功能:
(1)通过设置映射文件,使得用户信息属性延迟加载。
(2)通过添加二级缓存插件、设置映射文件,使得用户信息能够存放于二级缓存中。

第 8 章 Spring 原理与应用

本阶段目标

完成本章内容后,你将能够:
◆ 使用 Spring 管理 Bean 对象。

本阶段给出的步骤全面详细,请学员按照给出的上机步骤独立完成上机练习,以达到要求的学习目标。请认真完成下列步骤。

8.1 指导

本次上机完成的任务是在上一章上机完成的代码基础上,继续完成"我的宝贝儿"项目,要求如下:
(1)完成宠物信息查询的分页显示功能。
(2)完成宠物日志信息发表的时间类型和字符串类型相互转换的功能。
(3)使用 Spring 管理 Bean。
训练的技能点如下:
(1)使用 Spring 管理 Bean 的创建及维护。
(2)使用 Spring 管理程序中的依赖关系。

8.2 练习

分阶段完成以下任务:
● 阶段 1:创建分页显示的实用类 PageResult.java。
训练要点:使用 Spring 管理及维护 Bean 对象。
需求说明:
系统中,分页显示数据的业务逻辑是极其常见也是非常实用的功能,几乎所有的 Web 应用中都要使用分页逻辑。因此,在"电子宠物网站"中也加入了分页显示的功能。本章就分页显示实用类展开描述,并演示如何使用 Spring 管理和维护该类的实例对象。现在需要我们使用 Spring 的依赖注入管理 Biz 层对 DAO 层的依赖关系。

本阶段任务是创建实用类 PageResult.java，使用 Spring 组装 PageResult 类的实例并测试。

实现思路及关键代码如下：

我们首先要给项目增加 Spring 支持，这样，我们就可以在 Action 代码中以如下方式获得 Biz 的实例了。Spring 配置文件代码模板如示例代码 8-1 所示。

示例代码 8-1：Spring 配置文件 applicationContext.xml 代码模板

```xml
<?xml version="1.0" encoding="UTF-8"?>
<beans
    xmlns="http://www.springframework.org/schema/beans"
    xmlns:xsi="http://www.w3.org/2001/XMLSchema-instance"
    xmlns:p="http://www.springframework.org/schema/p"
    xsi:schemaLocation="http://www.springframework.org/schema/beans   http://www.springframework.org/schema/beans/spring-beans-3.1.xsd">



</beans>
```

使用 Spring 管理 PageResult 的实例步骤如下所示。

（1）PageResult.java 的代码如示例代码 8-2 所示。

示例代码 8-2：PageResult.java 的代码

```java
package com.xtgj.epet.comm;

import java.util.ArrayList;
import java.util.List;

/*
 *
 * @author 迅腾国际
 */
public class PageResult {
    private List list = new ArrayList(); // 查询结果
    private int pageNo = 1; // 实际页号
    private int pageSize = 15; // 每页记录数
    private int recTotal = 0; // 总记录数

    public PageResult() {
```

```java
            System.out.println("this is the constructor!");
    }

    public List getList() {
        return list;
    }

    public void setList(List list) {
        this.list = list;
    }

    public int getPageNo() {
        return pageNo;
    }

    public void setPageNo(int pageNo) {
        this.pageNo = pageNo;
    }

    public int getPageSize() {
        return (0 == pageSize) ? 10 : pageSize;
    }

    public void setPageSize(int pageSize) {
        this.pageSize = pageSize;
    }

    public int getRecTotal() {
        return recTotal;
    }

    public void setRecTotal(int recTotal) {
        this.recTotal = recTotal;
    }

    public int getPageTotal() {
        int ret = (this.getRecTotal() - 1) / this.getPageSize() + 1;
        ret = (ret < 1) ? 1 : ret;
```

```java
            return ret;
    }


    public int getFirstRec() {
            int ret = (this.getPageNo() - 1) * this.getPageSize();// + 1;
            ret = (ret < 1) ? 0 : ret;
            return ret;
    }

}
```

(2)使用 Spring 容器实例化 PageResult 的实例,Spring 配置文件如示例代码 8-3 所示。

示例代码 8-3:Spring 配置文件中添加对 PageResult 的实例管理

```xml
<?xml version="1.0" encoding="UTF-8"?>
<beans
    xmlns="http://www.springframework.org/schema/beans"
    xmlns:xsi="http://www.w3.org/2001/XMLSchema-instance"
    xmlns:p="http://www.springframework.org/schema/p"
    xsi:schemaLocation="http://www.springframework.org/schema/beans   http://www.springframework.org/schema/beans/spring-beans-3.1.xsd">

    <bean id="rageResult" class="com.xtgj.epet.comm.PageResult"
            scope="singleton">
    </bean>

</beans>
```

(3)编写测试用例,如示例代码 8-4 所示。

示例代码 8-4:测试用例 SpringTest.java 代码

```java
package com.xtgj.test;

import org.junit.After;
import org.junit.Before;
import org.junit.Test;
import org.springframework.context.ApplicationContext;
import org.springframework.context.support.ClassPathXmlApplicationContext;
```

```java
import com.xtgj.epet.comm.PageResult;

public class SpringTest {

    ApplicationContext context = null;

    @Before
    public void setUp() throws Exception {
        context = new ClassPathXmlApplicationContext("applicationContext.xml");
    }
    @After
    public void tearDown() throws Exception {
    }
    @Test
    public void testPage() throws Exception {
        PageResult pr = (PageResult) context.getBean("rageResult");
        PageResult pr1 = (PageResult) context.getBean("rageResult");
        System.out.println(" 页号:"+pr.getPageNo());
        System.out.println(" 页号:"+pr1.getPageNo());
    }
}
```

运行程序，显示如图 8-1 所示的运行结果。

图 8-1　当 bean 的 scope 属性值是 "singleton" 时的运行结果

修改 Spring 容器实例化 PageResult 的方式，如示例代码 8-5 所示。

示例代码 8-5：Spring 容器实例化 PageResult

```xml
<bean id="rageResult" class="com.xtgj.epet.comm.PageResult"
    scope="prototype">
</bean>
```

重新运行程序，显示如图 8-2 所示的运行结果。

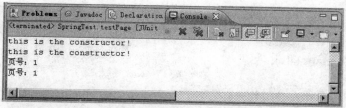

图 8-2 当 biean 的 scope 属性值是 "prototype" 时的运行结果

- 阶段 2：使用 Spring 管理实用类 Util.java。

与阶段 1 类似，组装 Util 相关的程序组件。Util 类如示例代码 8-6 所示。

示例代码 8-6：测试用例 SpringTest.java 代码

```java
package com.xtgj.epet.comm;

import java.text.ParseException;
import java.text.SimpleDateFormat;
import java.util.Calendar;
import java.util.Date;

public class Util {
    private static SimpleDateFormat df = new SimpleDateFormat("yyyy-MM-dd");
    private static SimpleDateFormat df2 = new SimpleDateFormat(
            "yyyy-MM-dd HH:mm:ss");
    private static SimpleDateFormat dfCN = new SimpleDateFormat("yyyy 年 MM 月 dd 日 ");

    /**
     * @param s
     * @return
     */
    public static boolean isNullOrBlank(final String s) {
        if (s == null) {
            return true;
        }
        if (s.trim().equals("")) {
```

```java
                    return true;
                }
                return false;
        }

        /**
         * @param s
         * @return
         */
        public static boolean isNullOrBlank(final Short s) {
                if (s == null || s.intValue() == -1) {
                        return true;
                }
                return false;
        }

        public static Date parseDate(String date) {
                Date ret = null;
                try {
                        ret = df.parse(date);
                } catch (ParseException e) {
                        ret = null;
                }
                return ret;
        }

        public static String formatDate(Date date) {
                String ret = "";
                try {
                        ret = df.format(date);
                } catch (Exception e) {
                        ret = "";
                }
                return ret;
        }

        public static String formatDateCN(Date date) {
                String ret = "";
```

```java
            try {
                ret = dfCN.format(date);
            } catch (Exception e) {
                ret = "";
            }
            return ret;
        }

        public static String formatTime(Date date) {
            String ret = "";
            try {
                ret = df2.format(date);
            } catch (Exception e) {
                ret = "";
            }
            return ret;
        }

        public static String getWeekNameCN(Date date) {
            String ret = "";
            try {
                ret = " 星期 ";
                Calendar cal = Calendar.getInstance();
                cal.setTime(date);
                int w = cal.get(Calendar.DAY_OF_WEEK);
                String ws = " 日一二三四五六 ";
                ret += ws.charAt(w - 1);
            } catch (Exception e) {
                ret = "";
            }
            return ret;
        }

        public static boolean isNotNullOrEmpty(String str) {
            if (null == str) {
                return false;
            }
            if (str.trim().equals("")) {
```

```
                    return false;
            }
            return true;
    }

    public static boolean isNotNullOrEmpty(Integer i) {
            if (null == i) {
                    return false;
            }
            if (i.intValue() == -1) {
                    return false;
            }
            return true;
    }
}
```

与阶段 1 类似，利用 Spring 容器管理和维护该类的实例并测试。

8.3 作业

1. 简述 Bean 的实例化方式（三种）。
2. 简述 Bean 的 Scope 属性值有哪几种，分别表示什么范围。
3. 简述 Bean 的生命周期。
4. 简述 Bean 的手工装配和自动装配形式。

第9章 Spring DI 详解

本阶段目标

完成本章内容后,你将能够:
◇ 使用 Spring 管理业务逻辑层对 DAO 层的依赖关系。

本阶段给出的步骤全面详细,请学员按照给出的上机步骤独立完成上机练习,以达到要求的学习目标。请认真完成下列步骤。

9.1 指导

本次上机完成的任务是在上一章上机完成的代码基础上,继续完成"我的宝贝儿"项目,要求如下:
(1)使用 Spring 管理业务逻辑层对 DAO 层的依赖关系。
(2)使用 Spring 配置数据源。
(3)使用 Spring 配置事务。
训练的技能点如下:
使用 Spring 管理程序中的依赖关系。

9.2 练习

分阶段完成以下任务:
● 阶段1:使用 Spring 依赖注入管理 Diary 相关的依赖。
训练要点:使用 Spring 依赖注入管理程序中的依赖关系。
需求说明:
系统中,业务逻辑层组件(如 DiaryBiz)对 DAO 层组件存在依赖。现有的代码将这些依赖关系硬编码在系统中,例如:"private PetDiaryDAO diaryDAO= new PetDiaryDAO();"。这样,当需要将程序重新组装(如改用别的 DAO 实现类)时,就需要修改代码,重新编译。这在程序已经发布的情况下是不方便做到的。现在需要我们使用 Spring 的依赖注入管理 Biz 层对

DAO 层的依赖关系。

本阶段任务是使用 Spring 依赖注入重新组装 DiaryBiz 和 PetDiaryDAO 相关的类。

实现思路及关键代码如下：

我们首先要给项目增加 Spring 支持，这样，我们就可以在 Action 代码中以如下方式获得 Biz 的实例了。Spring 配置文件如示例代码 9-1 所示。

示例代码 9-1：Spring 配置文件 applicationContext.xml 代码

```xml
<?xml version="1.0" encoding="UTF-8"?>
<beans xmlns="http://www.springframework.org/schema/beans"
    xmlns:xsi="http://www.w3.org/2001/XMLSchema-instance" xmlns:context="http://www.springframework.org/schema/context"
    xmlns:aop="http://www.springframework.org/schema/aop" xmlns:tx="http://www.springframework.org/schema/tx"
    xsi:schemaLocation="http://www.springframework.org/schema/beans
        http://www.springframework.org/schema/beans/spring-beans-3.1xsd
        http://www.springframework.org/schema/context
        http://www.springframework.org/schema/context/spring-context-3.1xsd
        http://www.springframework.org/schema/aop http://www.springframework.org/schema/aop/spring-aop-3.1.xsd
        http://www.springframework.org/schema/tx http://www.springframework.org/schema/tx/spring-tx-3.1.xsd">

    <!-- 加载数据源配置信息 -->
    <bean id="placeholderConfig"
        class="org.springframework.beans.factory.config.PropertyPlaceholderConfigurer">
        <property name="location">
            <value>classpath:datasource.properties
            </value>
        </property>
    </bean>
    <!-- 定义数据源 -->
    <bean id="dataSource" class="com.mchange.v2.c3p0.ComboPooledDataSource"
        destroy-method="close" dependency-check="none">
        <property name="driverClass">
            <value>${datasource.driverClassName}
            </value>
        </property>
        <property name="jdbcUrl">
```

```xml
            <value>${datasource.url}</value>
        </property>
        <property name="user">
            <value>${datasource.username}</value>
        </property>
        <property name="password">
            <value>${datasource.password}</value>
        </property>
        <property name="acquireIncrement">
            <value>${c3p0.acquireIncrement}</value>
        </property>
        <property name="initialPoolSize">
            <value>${c3p0.initialPoolSize}</value>
        </property>
        <property name="minPoolSize">
            <value>${c3p0.minPoolSize}</value>
        </property>
        <property name="maxPoolSize">
            <value>${c3p0.maxPoolSize}</value>
        </property>
        <property name="maxIdleTime">
            <value>${c3p0.maxIdleTime}</value>
        </property>
        <property name="idleConnectionTestPeriod">
            <value>${c3p0.idleConnectionTestPeriod}
            </value>
        </property>
        <property name="maxStatements">
            <value>${c3p0.maxStatements}</value>
        </property>
        <property name="numHelperThreads">
            <value>${c3p0.numHelperThreads}</value>
        </property>
</bean>
<!-- 定义 sessionFactory -->
<bean id="sessionFactory"
        class="org.springframework.orm.hibernate3.LocalSessionFactoryBean">
    <property name="dataSource">
```

```xml
                    <ref local="dataSource" />
                </property>
                <property name="mappingResources">
                    <list>
                        <value>com/xtgj/epet/entity/PetInfo.hbm.xml</value>
                        <value>com/xtgj/epet/entity/PetDiary.hbm.xml</value>
                    </list>
                </property>
                <property name="hibernateProperties">
                    <props>
                        <prop key="hibernate.dialect">${hibernate.dialect}</prop>
                        <prop key="hibernate.show_sql">${hibernate.show_sql}</prop>
                        <prop key="hibernate.jdbc.fetch_size">${hibernate.jdbc.fetch_size}</prop>
                        <prop key="hibernate.jdbc.batch_size">${hibernate.jdbc.batch_size}</prop>
                        <prop key="hibernate.cglib.use_reflection_optimizer">false</prop>
                        <prop key="hibernate.connection.release_mode">${hibernate.connection.release_mode}</prop>
                    </props>
                </property>
    </bean>
    <!-- 定义事务管理器 -->
    <bean id="myTransactionManager"
        class="org.springframework.orm.hibernate3.HibernateTransactionManager">
        <property name="sessionFactory">
            <ref local="sessionFactory" />
        </property>
    </bean>
    <!-- 声明式事务代理模板 -->
```

```xml
    <bean id="txProxyTemplate" abstract="true"
            class="org.springframework.transaction.interceptor.TransactionProxyFactoryBean">
        <property name="transactionManager">
            <ref bean="myTransactionManager" />
        </property>
        <property name="transactionAttributes">
            <props>
                <prop key="add*">PROPAGATION_REQUIRED</prop>
                <prop key="save*">PROPAGATION_REQUIRED</prop>
                <prop key="delete*">PROPAGATION_REQUIRED</prop>
                <prop key="update*">PROPAGATION_REQUIRED</prop>
                <prop key="do*">PROPAGATION_REQUIRED</prop>
                <prop key="*">PROPAGATION_REQUIRED,readOnly
                </prop>
            </props>
        </property>
    </bean>
    <!--
        省略部分代码
    -->
</beans>
```

使用 Spring 重新组装 Diary 相关类依赖关系的步骤如下：

（1）去掉 DiaryBiz 中直接实例化 PetDiaryDAO 实例的代码，如示例代码 9-2 所示。

示例代码 9-2：DiaryBiz 代码

```java
public class DiaryBiz {
    private PetDiaryDAO diaryDAO = null;

    public PetDiaryDAO getDiaryDAO() {
        return diaryDAO;
    }

    public void setDiaryDAO(PetDiaryDAO diaryDAO) {
        this.diaryDAO = diaryDAO;
    }
    ……
}
```

修改为"private PetDiaryDAO diaryDAO = null;",然后增加 setter 方法,为注入做准备。

(2)修改 DiaryActon 中直接实例化 DiaryBiz 的代码,获得由 Spring 创建的 Biz 类的实例。

示例代码 9-3:DiaryActon 代码

```java
public class DiaryAction {
    private DiaryBiz diaryBiz = null;
    private PetBiz petBiz = null;

    public DiaryBiz getDiaryBiz() {
            return diaryBiz;
    }

    public PetBiz getPetBiz() {
            return petBiz;
    }

    public void setPetBiz(PetBiz petBiz) {
            this.petBiz = petBiz;
    }

    public void setDiaryBiz(DiaryBiz diaryBiz) {
            this.diaryBiz = diaryBiz;
    }
//……
// 省略部分代码
}
```

(3)在 Spring 配置文件 applicationContext.xml 中配置,如示例代码 9-4 所示。

示例代码 9-4:applicationContext.xml 代码

```xml
<!--
     省略部分代码
-->
 <!--
     从这里开始配置。
          共四步:DAO -> Target -> Biz -> Action 。
-->
 <!-- 1. DAO -->
 <bean id="diaryDAO" class="com.xtgj.epet.dao.PetDiaryDAO">
```

```xml
            <property name="sessionFactory">
                <ref local="sessionFactory" />
            </property>
    </bean>
    <!-- 2. Target -->
    <bean id="diaryTarget" class="com.xtgj.epet.biz.DiaryBiz">
        <property name="diaryDAO">
            <ref local="diaryDAO" />
        </property>
    </bean>
    <!-- 3. Biz -->
    <bean id="diaryBiz" parent="txProxyTemplate">
        <property name="target">
            <ref bean="diaryTarget" />
        </property>
    </bean>


    <!-- 4. Action
    name 属性要和 struts.xml 中相应 Action 中 class 属性一致。
    -->
    <bean name="diary" class="com.xtgj.epet.web.action.DiaryAction">
        <property name="diaryBiz">
            <ref bean="diaryBiz" />
        </property>
        <property name="petBiz">
            <ref bean="petBiz" />
        </property>
    </bean>
    <!--
        省略部分代码
    -->
```

重新运行程序(如查看日记明细页,验证系统组装成功)。

● 阶段 2:使用 Spring 依赖注入管理 Pet 相关的依赖。

与阶段 1 类似,组装 Pet 相关的程序组件间的依赖。

之前 PetAction 中有在方法内部实例化 PetInfoBiz 实现类的,需要都改成通过调用成员变量 petInfoBiz 来调用业务方法。

9.3 作业

使用 Spring 依赖注入改造电子宠物网站中的用户管理程序。

第 10 章　Spring AOP

本阶段目标

完成本章内容后,你将能够:
✧ 使用 Spring 管理业务逻辑层对 DAO 层的依赖关系。
✧ 使用 Spring 配置事务。

本阶段给出的步骤全面详细,请学员按照给出的上机步骤独立完成上机练习,以达到要求的学习目标。请认真完成下列步骤。

10.1　指导

本次上机的任务是在上一章上机完成的代码基础上,继续完成"我的宝贝儿"项目,要求如下:
(1)使用 Spring 管理业务逻辑层对 DAO 层的依赖关系。
(2)使用 Spring AOP 处理特殊需求。
训练的技能点如下:
使用 AOP 灵活地实现系统需求。

10.2　练习

分阶段完成以下任务:
● 阶段 1:使用 AOP 的方式增加奖励功能。
在很多时候,我们不能随意修改业务逻辑实现的代码。有时是因为客户方严格的 IT 管理制度,有时是由于我们无法承担过高的修改成本,有时是为了防范引入不可预期缺陷的风险,有时是因为在升级项目中我们无法获得前期系统的源代码,有时是因为业务逻辑封装在 EJB 中,我们团队中没有掌握相关技术的人员,而业务的需要又是灵活多变的,唯一不变的是变化本身。

我们无时无刻不在努力寻找既稳妥又不失灵活地解决这一冲突的方法。AOP 给我们提供了新的思路。

现有需求如下:

(1) 我们的电子宠物网站会组织一些活动,在特定的时间内对特定的用户操作奖励积分。

(2) 需要在不影响系统正常运行的前提下实现一些功能,要求不能修改系统原有代码,包含 Action、Biz 和 DAO 的代码。

(3) 活动结束时,能方便地把功能去掉,不在系统中遗留垃圾代码。

(4) 现在实现的活动内容:用户喂养宠物则额外赠加宠物体力(petStrength)积分,不满 200 的,增加到 200;满 200 的,再增加 50。

实现思路及关键代码如下:

我们可以拦截 com.xtgj.epet.biz.doFeed(Long petId) 方法的调用来实现这一功能。根据业务需要选择使用"后置通知"。

(1) 首先,创建通知代码"LotteryAdvice.java",需要实现 AfterReturningAdvice 接口。从 afterReturning 方法的传入参数中,我们可以获得调用 doFeed 方法的参数(即宠物编号),然后调用 PetInfoDAO 获得宠物的信息。根据惯例,petInfoDAO 也采用注入的方式实例化。

(2) 然后,在 Spring 配置文件 applicationContext.xml 中完成配置。

LotteryAdvice.java 代码如示例代码 10-1 所示。

示例代码 10-1:LotteryAdvice.java

```java
package com.xtgj.epet.advice;

import java.lang.reflect.Method;

import org.springframework.aop.AfterReturningAdvice;

import com.xtgj.epet.dao.PetInfoDAO;
import com.xtgj.epet.entity.PetInfo;

public class LotteryAdvice implements AfterReturningAdvice {
    private PetInfoDAO petInfoDAO = null;

    public void afterReturning(Object returnValue, Method method,
                    Object[] args, Object target) throws Throwable {
        if (method.getName().equals("doFeed")) {
            Long petId = (Long) args[0];
            PetInfo petInfo = this.petInfoDAO.load(petId.toString());
            if (petInfo.getPetStrength() < 200) {
                petInfo.setPetStrength(200);
                this.petInfoDAO.update(petInfo);
```

```
                                System.out.println(petInfo.getPetName() + " 吃到了菠
菜,体力增长到 200！");
                    } else {
                            petInfo.setPetStrength(petInfo.getPetStrength() + 50);
                            this.petInfoDAO.update(petInfo);
                            System.out.println(petInfo.getPetName() + " 吃到了菠
菜,体力增加 50！");
                    }
            }
        }

        public void setPetInfoDAO(PetInfoDAO petInfoDAO) {
            this.petInfoDAO = petInfoDAO;
        }
    }
```

在 Spring 配置文件中添加如下配置,如示例代码 10-2 所示。

示例代码 10-2：Spring 配置文件中添加配置
```xml
    <bean id="petDiaryBiz"
            class="com.xtgj.epet.biz.impl.PetDiaryBizImpl">
            <property name="petDiaryDAO" ref="petDiaryDAO" />
    </bean>

    <bean id="lotteryAdvice"
            class="com.xtgj.epet.biz.LotteryAdvice">
            <property name="petInfoDAO" ref="petInfoDAO" />
    </bean>
    <bean id="petInfoBizTarget"
            class="com.xtgj.epet.biz.impl.PetInfoBizImpl">
            <property name="petInfoDAO" ref="petInfoDAO" />
    </bean>
    <bean id="petInfoBiz"
            class="org.springframework.aop.framework.ProxyFactoryBean">
            <property name="proxyInterfaces">
                    <value>com.xtgj.epet.biz.PetInfoBiz</value>
            </property>
            <property name="interceptorNames">
                    <list>
```

```xml
                    <value>lotteryAdvice</value>
                </list>
            </property>
            <property name="target" ref="petInfoBizTarget" />
</bean>
```

这样，用户喂养宠物时，便会增加宠物的体力值。此时，我们并没有改变业务逻辑代码的原始状态，而是仅仅通过配置方面和通知，完成了对 doFeed 方法的拦截和外部逻辑代码的注入。

- 阶段 2：使用 AOP 的方式增加日记记分功能。

与阶段 1 类似，以 AOP 的方式实现如下功能。

用户写宠物日记，则增加对应宠物的聪明值（PetCute）10 分，爱心值（PetLove）10 分。

参考步骤如下：

(1) 创建通知代码 DiaryAdvice（使用前置通知）。
(2) 在 Spring 配置文件中配置。

10.3 作业

1. 使用 AOP 方式实现下列功能：执行用户编辑时检查用户状态，如果用户状态为"已禁用"（flag 为 -1），则修改为"正常"（flag 为 0）。

2. 某销售系统销售奖励的计算方法是：当销售商家完成月（或季度）任务就可以获得销售奖励，对于不同级别的销售商，任务和奖励比率是不同的。现有系统中，这些数据在 properties 文件中配置。

销售商	月任务	月奖励比率	季度任务	季度奖励比率
区域代理	5000	3%	20000	5%
一级渠道商	1000	5%	5000	7%
零售商	100	7%	500	9%

现需要对系统升级，使这些数据可以在 Spring 配置文件中配置，注入系统的相应类别实例中。销售奖励实体类代码如示例代码 10-3 所示。

示例代码 10-3：销售奖励实体类 PerformanceConfig

```
public class PerformanceConfig {
    private Long monthTask;// 月任务
    private Double monthPrize;// 月奖励比率
    private Long quarterTask;// 季度任务
    private Double quarterPrize;// 季度奖励比率
```

```java
    public Long getMonthTask() {
        return monthTask;
    }

    public void setMonthTask(Long monthTask) {
        this.monthTask = monthTask;
    }

    public Double getMonthPrize() {
        return monthPrize;
    }

    public void setMonthPrize(Double monthPrize) {
        this.monthPrize = monthPrize;
    }

    public Long getQuarterTask() {
        return quarterTask;
    }

    public void setQuarterTask(Long quarterTask) {
        this.quarterTask = quarterTask;
    }

    public Double getQuarterPrize() {
        return quarterPrize;
    }

    public void setQuarterPrize(Double quarterPrize) {
        this.quarterPrize = quarterPrize;
    }
}
```

销售奖励业务逻辑类如示例代码 10-4 所示。

示例代码 10-4：销售奖励业务逻辑类代码 PerformanceBizImpl

```java
public class PerformanceBizImpl implements PerformanceBiz {
    // 销售奖励实体
    private PerformanceConfig performanceConfig = null;

    public PerformanceConfig getPerformanceConfig() {
        return performanceConfig;
    }

    public void setPerformanceConfig(PerformanceConfig performanceConfig) {
        this.performanceConfig = performanceConfig;
    }

    // 省略部分代码

}
```

计算绩效的 Action 如下，如所代码 10-5 所示。

示例代码 10-5：计算绩效的 Action

```java
public class PerformanceAction {
    // 区域代理绩效计算业务类
    private PerformanceBiz regionPerformanceBiz = null;
    // 一级渠道商绩效计算业务类
    private PerformanceBiz channelPerformanceBiz = null;
    // 零售店绩效计算业务类
    private PerformanceBiz salerPerformanceBiz = null;
    // Getter&Setter
    // ......
    // Action 方法
    // ......

}
```

请写出符合需求的 Spring 配置文件。

第 11 章 Spring 与 Struts、Hibernate 的集成

本阶段目标

完成本章内容后，你将能够：
◇ 掌握 Spring 与 Struts 2、Hibernate 的集成。

11.1 指导

本次上机的任务是使用 Spring 重新组装"我的宝贝儿"在线宠物网站。主要包含以下几方面：
(1) 将系统使用的 Spring 版本升级到 3.1。
(2) 使用 Spring 依赖管理后台代码。
(3) 使用 Spring 对 Hibernate 的支持重新组织持久化代码。
(4) 使用 Spring 2.5 风格声明式事务对业务逻辑方法提供事务支持。
本次上机任务训练的技能点包括以下几方面：
(1) 使用 Spring 管理程序中的依赖关系。
(2) 使用 Spring 对 Hibernate 的支持简化 Hibernate 编码。
(3) Spring 与 Struts 2 集成。
(4) 使用 Spring 配置声明式事务。

11.2 练习

分阶段完成以下任务：
● 阶段 1：使用 Spring 2.5 重新组装 pet 相关后台程序。
训练要点如下：
(1) 给项目添加 Spring 3.1 支持。

（2）使用 Spring 组织 JavaWeb 项目后台代码。
（3）Spring 与 Hibernate 的集成。
（4）Spring 和 Struts 2 集成。

下面给出本模块的实现思路及部分关键代码，如下所述：

第一步，整合开发时 Struts 2、Hibernate、Spring 需要的 jar 包括：

> Struts 2 核心安装包下的：
> struts2-core-2.x.x.jar：Struts 2 框架的核心类库
> xwork-core-2.x.x.jar：xwork 类库，Struts2 在其上构建
> ognl-2.6.x.jar：对象图导航语言（Object Graph Navigation Language），Struts 2 框架通过其读写对象的属性
> freemarker-2.3.x.jar：Struts 2 的 UI 标签的模板使用 FreeMarker 编写
> commons-fileupload-1.2.x.jar 文件上传组件，2.1.6 版本后需要加入此文件
> struts2-spring-plugin-2.x.x.jar：用于 struts2 集成 Spring 的插件
> hibernate 核心安装包下的（下载路径：http://www.hibernate.org/，点击"Hibernate Core"右边的"Downloads"）：
> hibernate4.jar
> lib\bytecode\cglib\hibernate-cglib-repack-3.1_3.jar
> lib\required*.jar
> hibernate 注解安装包下的（下载路径：www.hibernate.org，点击"Hibernate Annotations"右边的"Downloads"）：
> hibernate-annotations.jar
> lib\ejb3-persistence.jar、hibernate-commons-annotations.jar
> Hibernate 针对 JPA 的实现包（下载路径：www.hibernate.org，点击"Hibernate Entitymanager"右边的"Downloads"）：
> hibernate-entitymanager.jar
> lib\test\log4j.jar、slf4j-log4j12.jar
> Spring 安装包下的：
> dist\spring.jar
> lib\c3p0\c3p0-0.9.1.2.jar
> lib\aspectj\aspectjweaver.jar、
> aspectjrt.jar
> lib\cglib\cglib-nodep-2.1_3.jar
> lib\j2ee\common-annotations.jar
> lib\log4j\log4j-1.2.15.jar
> lib\jakarta-commons\commons-logging.jar
> 数据库驱动 jar

第二步，给 Spring 配置文件增加命名空间的声明，如示例代码 11-1 所示：

第 11 章　Spring 与 Struts、Hibernate 的集成

示例代码 11-1：给 Spring 配置文件增加命名空间的声明

```xml
<?xml version="1.0" encoding="UTF-8"?>
<beans xmlns="http://www.springframework.org/schema/beans"
    xmlns:xsi="http://www.w3.org/2001/XMLSchema-instance"
    xmlns:context="http://www.springframework.org/schema/context"
    xmlns:aop="http://www.springframework.org/schema/aop"
    xmlns:tx="http://www.springframework.org/schema/tx"
    xsi:schemaLocation="http://www.springframework.org/schema/beans
    http://www.springframework.org/schema/beans/spring-beans-3.1.xsd
    http://www.springframework.org/schema/context
    http://www.springframework.org/schema/context/spring-context-3.1.xsd
    http://www.springframework.org/schema/aop   http://www.springframework.org/schema/aop/spring-aop-3.1.xsd
    http://www.springframework.org/schema/tx http://www.springframework.org/schema/tx/spring-tx-2.5.xsd">

</beans>
```

第三步，修改 PetInfoDAO.java 代码，继承自 BaseDAO，BaseDAO 继承自 HibernateDaoSupport 实现。重新编写方法的实现代码。PetInfoDAO 代码如示例代码 11-2 所示。

示例代码 11-2：PetInfoDAO 代码

```java
package com.xtgj.epet.dao;
// 省略 import
/**
 * 数据访问对象
 * @see .PetInfo
 * @author 迅腾国际
 */
public class PetInfoDAO extends BaseDAO {
    public void list(String listType,PetInfo condition,PageResult pageResult)
    {
            String hql = "from PetInfo p where 1=1 ";
            if (null!=condition)
            {
                    if (Util.isNotNullOrEmpty(condition.getPetName()))
                    {
                            hql += "and p.petName like '%" + condition.getPetName() + "%' ";
```

```java
            }
            if (Util.isNotNullOrEmpty(condition.getPetType()))
            {
                hql += "and p.petType = " + condition.getPetType() + " ";
            }
            if (Util.isNotNullOrEmpty(condition.getPetOwnerName()))
            {
                hql += "and p.petOwnerName like '%" + condition.getPetOwnerName() +"%' ";
            }
        }
        if ("all".equals(listType) || null==listType)
        {
            hql += "order by p.petScore desc";
        }else if ("cute".equals(listType))
        {
            hql += "order by p.petCute desc";
        }else if ("love".equals(listType))
        {
            hql += "order by p.petLove desc";
        }else if ("strength".equals(listType))
        {
            hql += "order by p.petStrength desc";
        }
        super.list(hql, pageResult);
    }
    public PetInfo load(String id)
    {
        PetInfo ret = null;
        String hql = "from PetInfo p where p.petId = " + id;
        List list = super.list(hql);
        if(null!=list && list.size()==1)
        {
            ret = (PetInfo)list.get(0);
        }
        return ret;
    }
    public PetInfo load(Integer id)
```

```
                {
                        if (id == null)
                        {
                                return null;
                        }
                        PetInfo ret = this.load(id.intValue() + "");
                        return ret;
                }
        public List topList(String listType, int count) {
                String hql = "from PetInfo p ";

                if ("all".equals(listType))
                {
                        hql += "order by p.petScore desc";
                }else if ("cute".equals(listType))
                {
                        hql += "order by p.petCute desc";
                }else if ("love".equals(listType))
                {
                        hql += "order by p.petLove desc";
                }else if ("strength".equals(listType))
                {
                        hql += "order by p.petStrength desc";
                }
                Query query = this.getSession().createQuery(hql);
                query.setFirstResult(0);
                query.setMaxResults(count);
                List ret = query.list();
                return ret;
        }
}
```

第四步，在 Web 容器中使用 Listener 实例化 Spring 容器和配置 Struts 2，web.xml 代码片段示例代码如下：

示例代码 11-3：web.xml 配置文件

```
<!-- 指定 Spring 的配置文件，默认从 web 根目录寻找配置文件，我们可以通过
spring 提供的 classpath: 前缀指定从类路径下寻找 -->
<context-param>
```

```xml
        <param-name>contextConfigLocation</param-name>
        <param-value>classpath:applicationContext.xml</param-value>
    </context-param>
    <!-- 对 Spring 容器进行实例化 -->
    <listener>
        <listener-class>org.springframework.web.context.ContextLoaderListener</listener-class>
    </listener>
    <!-- 配置 Struts2-->
    <filter>
        <filter-name>struts2</filter-name>
        <filter-class>org.apache.struts2.dispatcher.ng.filter.StrutsPrepareAndExecuteFilter</filter-class>
    </filter>
    <filter-mapping>
        <filter-name>struts2</filter-name>
        <url-pattern>/*</url-pattern>
    </filter-mapping>
```

第五步，在 Struts 2 的配置文件 struts.xml 中配置 Spring 与 Struts 2 的集成。struts.xml 代码如示例代码 11-4 所示。

示例代码 11-4：struts.xml 代码

```xml
<?xml version="1.0" encoding="UTF-8" ?>
<!DOCTYPE struts PUBLIC
    "-//Apache Software Foundation//DTD Struts Configuration 2.0//EN"
    "http://struts.apache.org/dtds/struts-2.0.dtd">

<struts>
    <!-- 默认的视图主题 -->
    <constant name="struts.ui.theme" value="simple" />

    <constant name="struts.objectFactory" value="spring" />

    <package name="pet" namespace="/pet" extends="struts-default">
        <action name="pet_*" class="pet" method="{1}">
            <result name="index">/WEB-INF/jsp/pet/index.jsp
            </result>
            <result name="welcome" type="redirectAction">pet_index</result>
```

```xml
            <result name="baby">/WEB-INF/jsp/pet/baby.jsp
            </result>
            <result name="adopt">/WEB-INF/jsp/pet/adopt.jsp
            </result>
            <result name="doAdopt" type="redirectAction">
                    pet_toBaby?petForm.id=${petForm.id}</result>
            <result name="Training" type="redirectAction">
                    pet_toBaby?petForm.id=${petForm.id}
            </result>
            <result name="list">/WEB-INF/jsp/pet/list.jsp
            </result>
        </action>

    </package>

    <package name="diary" namespace="/diary" extends="struts-default">
        <action name="diary_*" class="diary" method="{1}">
            <result name="list">/WEB-INF/jsp/diary/list.jsp
            </result>
            <result name="diary">/WEB-INF/jsp/diary/diary.jsp
            </result>
            <result name="toBaby" type="redirectAction">
                    <param name="actionName">pet_toBaby?petForm.id=${petForm.id}</param>
                    <param name="namespace">/pet</param>
            </result>
            <result name="detail">/WEB-INF/jsp/diary/detail.jsp
            </result>
        </action>
    </package>
</struts>
```

第六步，修改 ActionBean 创建 Biz 类实例的代码，修改为通过 setter 方法注入，此举是为将 Biz 类实例注入 ActionBean 做准备。PetAction 代码如示例代码 11-5 所示。

示例代码 11-5：PetAction 代码

```java
package com.xtgj.epet.web.action;
// 省略 import

/*
 * PetAction
 * @author 迅腾国际
 */
public class PetAction{
    private PetBiz petBiz = null;
    private DiaryBiz diaryBiz = null;
    PetForm petForm;
    /*
     * 首页
     */
    public String index() {
        List all = this.getPetBiz().topList("all",10);
        ActionContext.getContext().put("all", all);
        List cute = this.getPetBiz().topList("cute",10);
        ActionContext.getContext().put("cute", cute);
        List strength = this.getPetBiz().topList("strength",10);
        ActionContext.getContext().put("strength", strength);
        List love = this.getPetBiz().topList("love",10);
        ActionContext.getContext().put("love", love);
        // 一个日记列表
        List newDiary = this.getDiaryBiz().topList("new",10);
        ActionContext.getContext().put("newDiary", newDiary);
        return "index";
    }

    /*
     * 登录
     * @throws IOException
     */
    public String doLogin(){
        boolean bLogin = this.getPetBiz().login(petForm);
        if (bLogin)
        {
```

```java
                        ActionContext.getContext().getSession().put("CURRENT_PET",
petForm.getCondition());
                        ActionContext.getContext().put("petForm.operate", "toBaby");
                        ActionContext.getContext().put("petForm.id", petForm.getCondi-
tion().getPetId().intValue());
                        return "welcome";
            }
            else
            {
                        ActionContext.getContext().put("petForm.operate", "index");
                        ActionContext.getContext().put("petForm.sign", "login_error");
                        return "welcome";

            }
}
/*
 * 注销
 */
public String doLogout(){
            ActionContext.getContext().getSession().remove("CURRENT_PET");
            return "welcome";
}
/*
 * 转到宠物页面
 */
public String toBaby() {
            PetInfo petInfo = this.getPetBiz().load(petId);
            petForm.setItem(petInfo);
            int month = 0;
            int year = 0;
            if (0==month)
            {
                        year = Calendar.getInstance().get(Calendar.YEAR);
                        month = Calendar.getInstance().get(Calendar.MONTH) + 1;
            }
            HashMap diaryList = this.getDiaryBiz().list(petId,year,month);
            ActionContext.getContext().put("diaryList", diaryList);
            ActionContext.getContext().put("petForm", petForm);
```

```java
        return "baby";
    }
    /*
     * 转到领养页面
     */
    public String toAdopt() {
        return "adopt";
    }
    /*
     * 执行领养操作
     * @throws IOException
     */
    public String doAdopt() {
        this.getPetBiz().doAdopt(petForm.getItem());
        ActionContext.getContext().put("petForm.operate", "toBaby");
        ActionContext.getContext().put("petForm.sign", "new");
        ActionContext.getContext().put("petForm.id",    petForm.getItem().getPetId());
        
        return "doAdopt";
    }
    
    /*
     * 训练
     * @throws IOException
     */
    public String doTraining(){
        this.getPetBiz().doTraining(petForm);
        ActionContext.getContext().put("petForm.id",    petForm.getItem().getPetId());
        return "Training";
    }
    
    /*
     * 列表
     */
    public String list() {
        this.getPetBiz().list(petForm.getListType(),petForm.getCondition(), petForm.getPageResult());
```

```java
                    ActionContext.getContext().put("petForm.pageResult.firstRec", petForm.getPageResult().getFirstRec());
            return "list";
    }

    public PetBiz getPetBiz() {
            return petBiz;
    }

    public void setPetBiz(PetBiz petBiz) {
            this.petBiz = petBiz;
    }

    public DiaryBiz getDiaryBiz() {
            return diaryBiz;
    }

    public void setDiaryBiz(DiaryBiz diaryBiz) {
            this.diaryBiz = diaryBiz;
    }

    public PetForm getPetForm() {
            return petForm;
    }

    public void setPetForm(PetForm petForm) {
            this.petForm = petForm;
    }

}
```

第七步，在 Spring 配置文件中组装程序。Spring 配置文件 applicationContext.xml 如示例代码 11-6 所示。

示例代码 11-6：applicationContext.xml 源代码

```xml
<!-- 加载数据源配置信息 -->
<bean id="placeholderConfig"
```

```xml
        class="org.springframework.beans.factory.config.PropertyPlaceholderConfigurer">
            <property name="location">
                <value>classpath:datasource.properties
                </value>
            </property>
    </bean>
    <!-- 定义数据源 -->
    <bean id="dataSource" class="com.mchange.v2.c3p0.ComboPooledDataSource"
        destroy-method="close" dependency-check="none">
        <property name="driverClass">
            <value>${datasource.driverClassName}
            </value>
        </property>
        <property name="jdbcUrl">
            <value>${datasource.url}</value>
        </property>
        <property name="user">
            <value>${datasource.username}</value>
        </property>
        <property name="password">
            <value>${datasource.password}</value>
        </property>
        <property name="acquireIncrement">
            <value>${c3p0.acquireIncrement}</value>
        </property>
        <property name="initialPoolSize">
            <value>${c3p0.initialPoolSize}</value>
        </property>
        <property name="minPoolSize">
            <value>${c3p0.minPoolSize}</value>
        </property>
        <property name="maxPoolSize">
            <value>${c3p0.maxPoolSize}</value>
        </property>
        <property name="maxIdleTime">
            <value>${c3p0.maxIdleTime}</value>
        </property>
```

```xml
            <property name="idleConnectionTestPeriod">
                <value>${c3p0.idleConnectionTestPeriod}
                </value>
            </property>
            <property name="maxStatements">
                <value>${c3p0.maxStatements}</value>
            </property>
            <property name="numHelperThreads">
                <value>${c3p0.numHelperThreads}</value>
            </property>
    </bean>
    <!-- 定义 sessionFactory -->
    <bean id="sessionFactory"
            class="org.springframework.orm.hibernate3.LocalSessionFactoryBean">
            <property name="dataSource">
                <ref local="dataSource" />
            </property>
            <property name="mappingResources">
                <list>
                        <value>com/xtgj/epet/entity/PetInfo.hbm.xml
                        </value>
                        <value>com/xtgj/epet/entity/PetDiary.hbm.xml
                        </value>
                </list>
            </property>
            <property name="hibernateProperties">
                <props>
                        <prop key="hibernate.dialect">${hibernate.dialect}</prop>
                        <prop key="hibernate.show_sql">${hibernate.show_sql}</prop>
                        <prop key="hibernate.jdbc.fetch_size">${hibernate.jdbc.fetch_size}
                        </prop>
                        <prop key="hibernate.jdbc.batch_size">${hibernate.jdbc.batch_size}
                        </prop>
```

```xml
                              <prop key="hibernate.cglib.use_reflection_optimizer">false</prop>
                              <prop key="hibernate.connection.release_mode">${hibernate.connection.release_mode}</prop>
                         </prop>
                    </props>
               </property>
     </bean>
     <!-- 定义事务管理器 -->
     ……

     <!--
          从这里开始配置。
                    共四步：DAO -> Target -> Biz -> Action 。
     -->
     <!-- 1. DAO -->
     <bean id="petDAO" class="com.xtgj.epet.dao.PetInfoDAO">
          <property name="sessionFactory">
               <ref local="sessionFactory" />
          </property>
     </bean>
     <bean id="diaryDAO" class="com.xtgj.epet.dao.PetDiaryDAO">
          <property name="sessionFactory">
               <ref local="sessionFactory" />
          </property>
     </bean>
     <!-- 2. Target -->
     <bean id="petTarget" class="com.xtgj.epet.biz.PetBiz">
          <property name="petDAO">
               <ref local="petDAO" />
          </property>
     </bean>
     <bean id="diaryTarget" class="com.xtgj.epet.biz.DiaryBiz">
          <property name="diaryDAO">
               <ref local="diaryDAO" />
          </property>
     </bean>
     <!-- 3. Biz -->
```

```xml
<bean id="petBiz" parent="txProxyTemplate">
    <property name="target">
        <ref bean="petTarget" />
    </property>
</bean>
<bean id="diaryBiz" parent="txProxyTemplate">
    <property name="target">
        <ref bean="diaryTarget" />
    </property>
</bean>

<!-- 4. Action
name 属性要和 struts.xml 中相应 Action 中 class 属性一致。
-->
<bean name="pet" class="com.xtgj.epet.web.action.PetAction">
    <property name="petBiz">
        <ref bean="petBiz" />
    </property>
    <property name="diaryBiz">
        <ref bean="diaryBiz" />
    </property>
</bean>
<bean name="diary" class="com.xtgj.epet.web.action.DiaryAction">
    <property name="diaryBiz">
        <ref bean="diaryBiz" />
    </property>
    <property name="petBiz">
        <ref bean="petBiz" />
    </property>
</bean>
```

- 阶段 2：使用 Spring 3.1 重新组装 diary 相关后台程序。

与阶段 1 类似，组装 diary 相关的程序组件间的依赖。

提示：

（1）修改 DiaryAction 代码，为注入做准备。

（2）修改 PetDiaryDAO 代码，继承自 HibernateSupport 实现。

（3）修改 struts.xml 中 DiaryAction 的配置。

（4）在 Spring 配置文件中组装。

- 阶段 3：使用 Spring 2.5 的方式给业务逻辑方法添加声明式事务支持。

下面给出本模块的实现思路及部分关键代码,如下所述:

声明式事务是 Spring AOP 的一个应用,声明式事务的配置也是以 AOP 的配置为基础的。Spring 2.5 提供了 tx 命名空间,用于在配置文件中声明通知等事务相关的配置。

首先,我们定义事务管理器(TransactionManager),然后定义事务通知,最后声明方面并将通知织入。参考解决方案和事务相关配置代码如示例代码 11-7 所示。

```xml
示例代码 11-7:事务相关配置
        <!-- 定义事务管理器 -->
        <bean id="myTransactionManager"
              class="org.springframework.orm.hibernate3.HibernateTransaction-Manager">
            <property name="sessionFactory">
                <ref local="sessionFactory" />
            </property>
        </bean>
        <!-- 声明式事务代理模板 -->
        <bean id="txProxyTemplate" abstract="true"
              class="org.springframework.transaction.interceptor.TransactionProxyFactoryBean">
            <property name="transactionManager">
                <ref bean="myTransactionManager" />
            </property>
            <property name="transactionAttributes">
                <props>
                    <prop key="add*">PROPAGATION_REQUIRED</prop>
                    <prop key="save*">PROPAGATION_REQUIRED</prop>
                    <prop key="delete*">PROPAGATION_REQUIRED</prop>
                    <prop key="update*">PROPAGATION_REQUIRED</prop>
                    <prop key="do*">PROPAGATION_REQUIRED</prop>
                    <prop key="*">PROPAGATION_REQUIRED,readOnly</prop>
                </props>
            </property>
        </bean>
```

● 阶段4:根据业务需求配置事务支持。

按如下需求配置事务:

在 Action Bean 中,以 to 开头的方法要求只读事务,以 do 开头的方法要求事务支持。

这里,我们只需修改 Spring 配置文件中的相关代码即可。

11.3 作业

1. 继续改造上一章作业完成的简单用户管理程序,使用 Spring 对 Hibernate 支持重新实现 DAO 层实现类。

2. 继续改造上一章作业完成的简单用户管理程序,使用 Spring 重新组装后台程序。

3. 继续改造上一章作业完成的简单用户管理程序,给业务逻辑层方法添加事务支持。